Oxford **Mathematics**

Primary Years Programme

Contents

OXFORD
UNIVERSITY PRESS
AUSTRALIA & NEW ZEALAND

Counting to 100

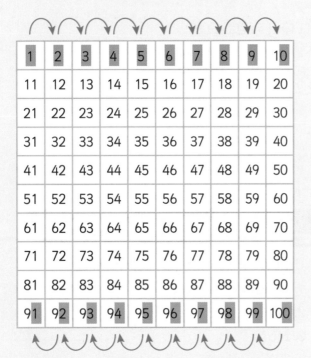

Forwards counting pattern by 1s

Backwards counting pattern by 1s

What is the pattern in the tens column?

Guided practice

1 What comes before and after ...?

a   22   23   24   b   [ ]   37   [ ]

c   [ ]   55   [ ]   d   [ ]   68   [ ]

e   [ ]   72   [ ]   f   [ ]   30   [ ]

OXFORD UNIVERSITY PRESS

Independent practice

1 Fill in the missing numbers.

a

26	27	28			31			

b

43		45				49		

c

66	67							

----

2 Fill in the missing numbers.

a

35　36　☐　☐　39　☐　☐　42　☐　☐　45

b

78　79　☐　☐　☐　83　☐　☐　86　☐　88

c

33　32　31　☐　☐　☐　☐　26　☐　☐　23

d

55　54　☐　☐　51　☐　☐　☐　☐　46　☐

Tens and ones

1 ten	1 ten and 2 ones	2 tens and 4 ones	3 tens and 7 ones
10	12	24	37

How many sticks are in each bundle? Do you need to count the bundled sticks every time?

Guided practice

1 How many ...

a

tens? ☐

ones? ☐

altogether? ☐

b

tens? ☐

ones? ☐

altogether? ☐

c

tens? ☐

ones? ☐

altogether? ☐

d

tens? ☐

ones? ☐

altogether? ☐

OXFORD UNIVERSITY PRESS

Independent practice

1 Group in 10s, then count.

a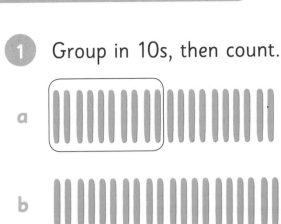

b

c

2 Match the pictures with the numbers.

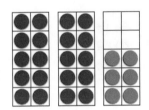 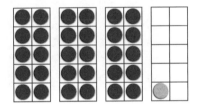 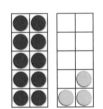 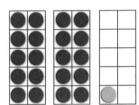

| 26 | 13 | 31 | 21 |

3 Draw counters to show 47.

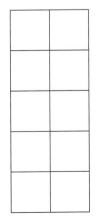 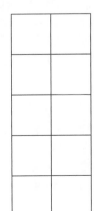 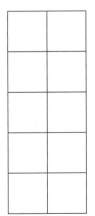 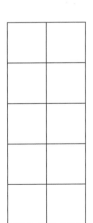 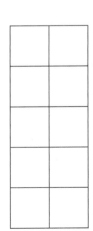

1 Who am I?

a I have 5 tens.

I have 9 ones.

I am ⬚⬚ .

b I have 1 ten.

I have 6 ones.

I am ⬚⬚ .

c I have 2 tens.

I am less than 21.

I am ⬚⬚ .

d I have 8 tens.

I am more than 88.

I am ⬚⬚ .

2 What is ...

a 2 more than 48? ⬚

b 2 less than 61? ⬚

c 1 more than 4 tens? ⬚

d 1 less than 3 tens? ⬚

OXFORD UNIVERSITY PRESS

Numbers can be shown with:

words	numerals	pictures
twenty-four	**24**	

> All compound numbers are written with a hyphen in them — twenty-four, thirty-three, ninety-nine.

Guided practice

1 Write the numerals.

a   twelve   ☐

b   twenty-eight   ☐

c   fifteen   ☐

d   fifty-three   ☐

e   fourteen   ☐

f   forty-five   ☐

2 Circle the correct way to write the numbers.

a	18	eighty	eighteen	eighty-one

b	46	fourty six	sixty-four	forty-six

1 Match the player with the shirt.

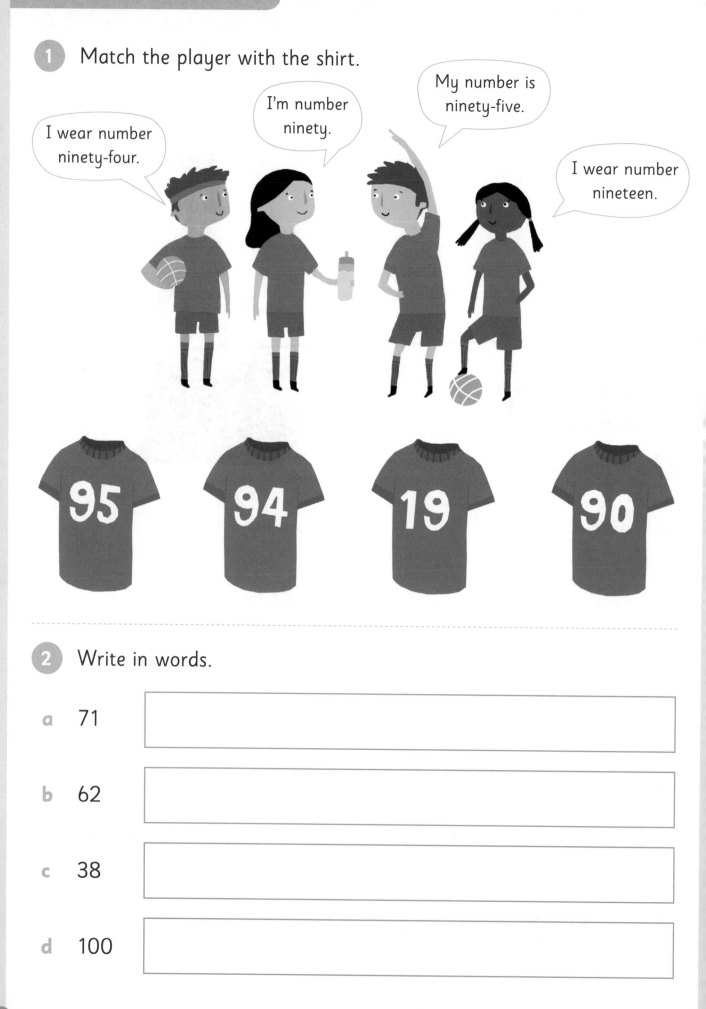

2 Write in words.

a 71

b 62

c 38

d 100

3 Match the words, pictures and numerals.

sixty-three

36

seventeen

23

twenty-two

63

twenty-three

42

thirty-six

17

forty-two

22

I wonder why "nine" doesn't change to make the word "ninety", but "five" changes to make "fifty"?

1 Write words and numerals for:

a

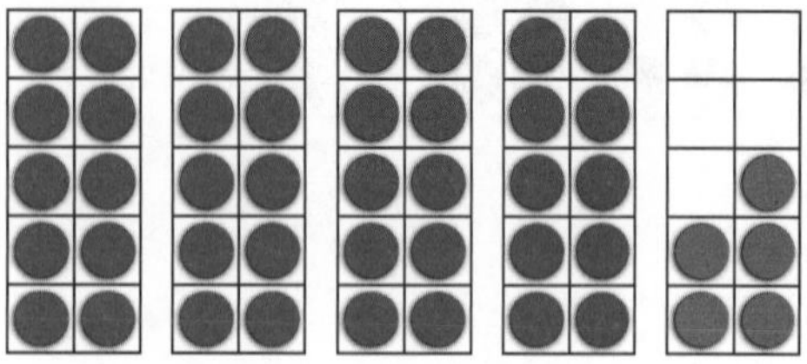

Words

_____

Numeral ☐

b

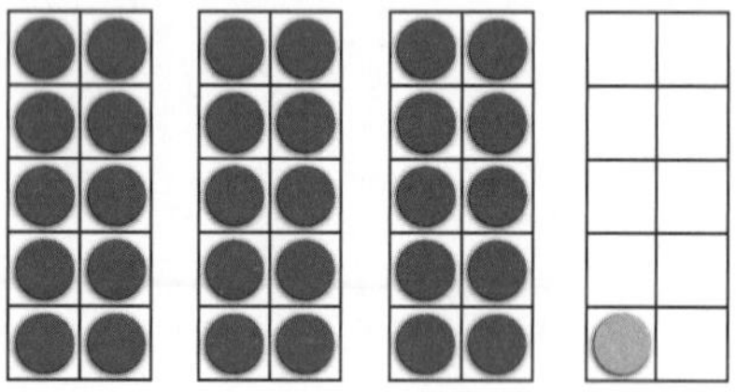

Words

_____

Numeral ☐

c

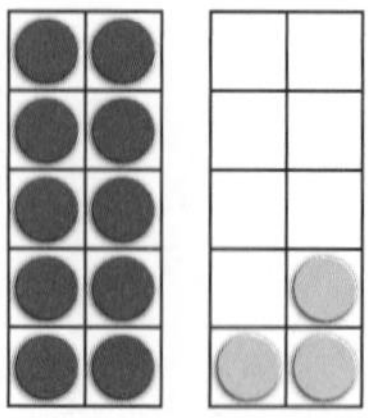

Words

_____

Numeral ☐

d

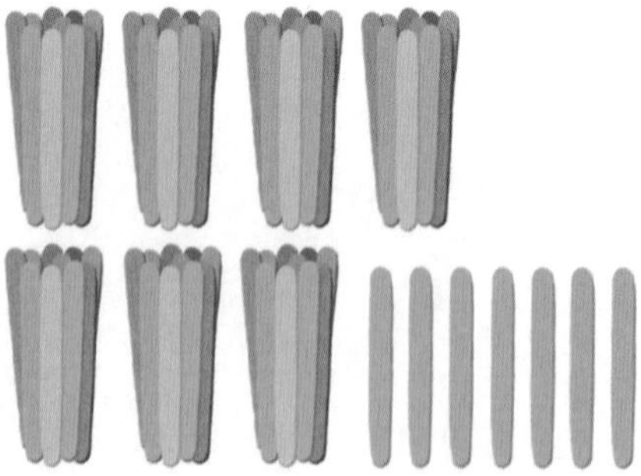

Words

_____

Numeral ☐

e

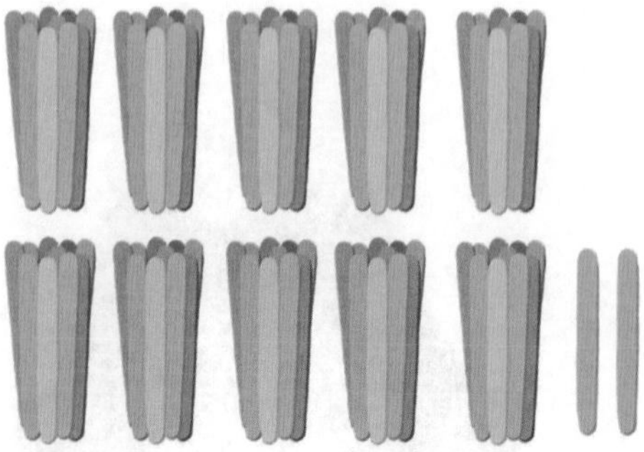

Words

_____

Numeral ☐

OXFORD UNIVERSITY PRESS

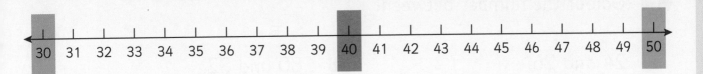

| 30 | 31 | 32 | 33 | 34 | 35 | 36 | 37 | 38 | 39 | 40 | 41 | 42 | 43 | 44 | 45 | 46 | 47 | 48 | 49 | 50 |

40 is bigger than **30**.

50 is bigger than **40**.

40 is smaller than **50**.

Look at the tens column first to work out which 2-digit number is bigger.

Guided practice

1 Colour the correct word.

| 20 | 21 | 22 | 23 | 24 | 25 | 26 | 27 | 28 | 29 | 30 | 31 | 32 | 33 | 34 | 35 | 36 | 37 | 38 | 39 | 40 |

a   30 is

bigger
smaller

than 20.

b   31 is

bigger
smaller

than 29.

2 Colour the correct word.

| 40 | 41 | 42 | 43 | 44 | 45 | 46 | 47 | 48 | 49 | 50 | 51 | 52 | 53 | 54 | 55 | 56 | 57 | 58 | 59 | 60 |

a   45 is

bigger
smaller

than 54.

b   48 is

bigger
smaller

than 52.

c   57 is

bigger
smaller

than 47.

d   50 is

bigger
smaller

than 46.

1 Colour the number between:

a 24 and 26.

| 23 | 25 | 27 |

b 80 and 82.

| 81 | 82 | 83 |

c 49 and 51.

| 40 | 48 | 50 |

d 77 and 80.

| 78 | 81 | 75 |

2 Match the numbers, pictures and words.

73 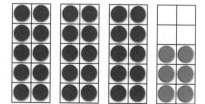 Less than 20

36 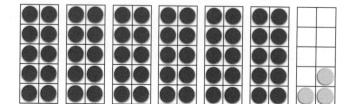 1 less than 37

63 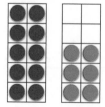 More than 70

16 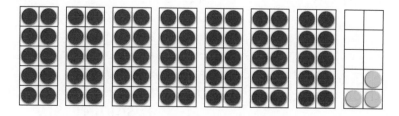 1 more than 62

OXFORD UNIVERSITY PRESS

3 Write the numbers in the correct places.

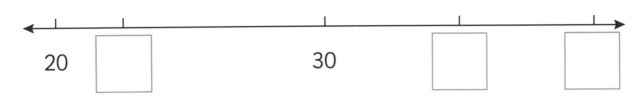

a 40   b 35   c 22

---

4 Write the numbers in the correct places.

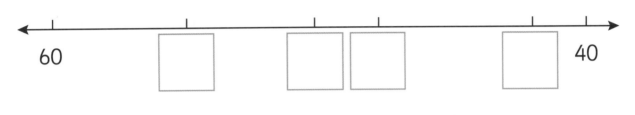

a 50   b 42   c 55   d 48

---

5 Write the numbers from **smallest** to **largest**.

13   67   113   48   37   52   84

---

6 Write the numbers from **largest** to **smallest**.

51   86   74   15   105   21   39

Are 2-digit or 3-digit numbers bigger? Why?

60	143	234
725	47	18	180

1 Write:

a the biggest number. ☐   **b** the smallest number. ☐

c the numbers with a 4 in the tens place. ☐ ☐

d the numbers smaller than 50. ☐ ☐

2 Write the numbers in the correct place.

0 ————————————————————————————— 50

a 40   **b** 25   **c** 10   **d** 3   **e** 38

3 Write from **smallest** to **largest**.

346   634   436   406   364   643

☐ ☐ ☐ ☐ ☐ ☐

13 and 4 is **17**

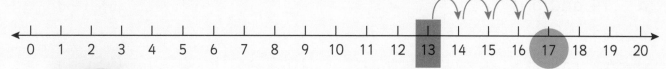

Start from the bigger number to count on — 13, 14, 15, 16, 17.

Why is it easier to count on from the **bigger** number?

Guided practice

1 Count on to find the answers.

a 9 and 3 is ☐.

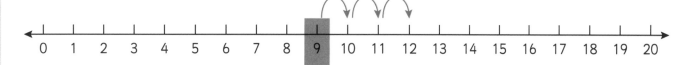

b 11 and 6 is ☐.

c 2 and 15 is ☐.

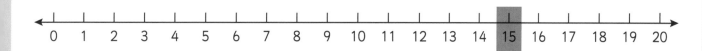

1 Circle the bigger number. Then count on.

a 14 and 5 is ☐ .

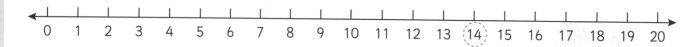

b 3 and 16 is ☐ .

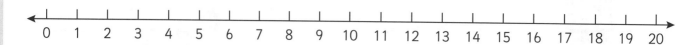

c 12 and 11 is ☐ .

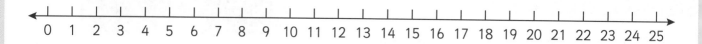

2 Show on the number line and solve.

a 10 and 4 is ☐ .

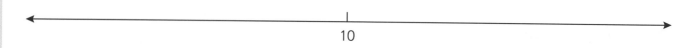

b 4 and 13 is ☐ .

c 11 and 6 is ☐ .

OXFORD UNIVERSITY PRESS

3

a   How many? ☐

b   Draw 6 more.

c   How many now? ☐

> Remember, you can start from the bigger number. You don't have to count them all again.

4

a   How many? ☐

b   Draw 8 more.

c   How many now? ☐

5   8 and 16

a   Draw the bigger number in **red**.

b   Draw the smaller number in **blue** to count on.

c   How many altogether? ☐

6   15 and 7

a   Draw the bigger number in **red**.

b   Draw more in **blue** to count on.

c   How many altogether? ☐

Extended practice

1 Count on from the bigger number.

a 23 and 9 is [ ] .

b 6 and 25 is [ ] .

c 4 and 31 is [ ] .

d 37 and 7 is [ ] .

e 32 and 5 is [ ] .

f 12 and 26 is [ ] .

2 Count on to find:

1	2	3	4	5	6	7	8	9	10
11	12	13	14	15	16	17	18	19	20
21	22	23	24	25	26	27	28	29	30
31	32	33	34	35	36	37	38	39	40
41	42	43	44	45	46	47	48	49	50
51	52	53	54	55	56	57	58	59	60
61	62	63	64	65	66	67	68	69	70
71	72	73	74	75	76	77	78	79	80
81	82	83	84	85	86	87	88	89	90
91	92	93	94	95	96	97	98	99	100

a 5 more than 42. [ ]

b 7 more than 53. [ ]

c 12 more than 65. [ ]

d 8 more than 86. [ ]

Partitioning means separating.

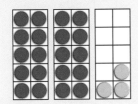

23 can be partitioned as:

10 and 13

20 and 3

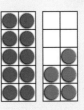 

15 and 8

Guided practice

How else could you partition 23?

1 Record how the numbers have been partitioned.

a is the same as and

7

b 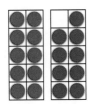 is the same as  and

19

c 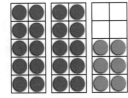 is the same as 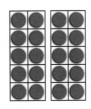 and

1 Draw counters to show the partitions. Then fill in the gaps.

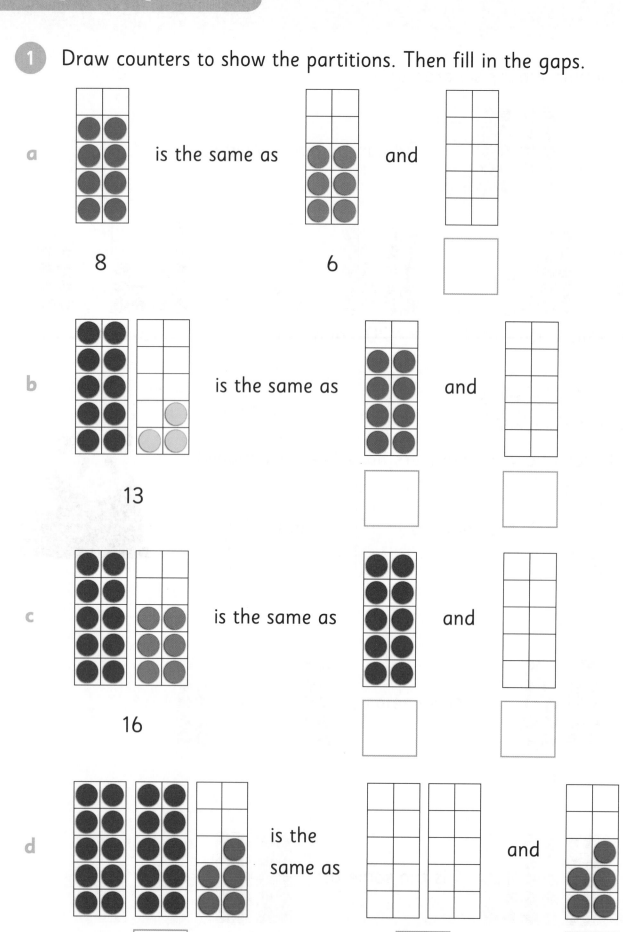

a   is the same as   and

8   6

b   is the same as   and

13

c   is the same as   and

16

d   is the same as   and

2 Partition each number into 2 parts.

a
7

6 and [ ]

b
14

[ ] and 7

c
10

6 and [ ]

d
17

7 and [ ]

e
14

10 and [ ]

f
28

[ ] and 8

g
25

5 and [ ]

h
29

19 and [ ]

When do you think it would be useful to partition numbers?

1 Partition each number 2 ways.

a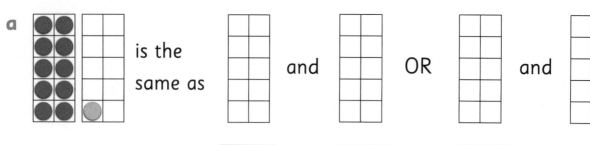

11 is the same as ☐ and ☐ OR ☐ and ☐

b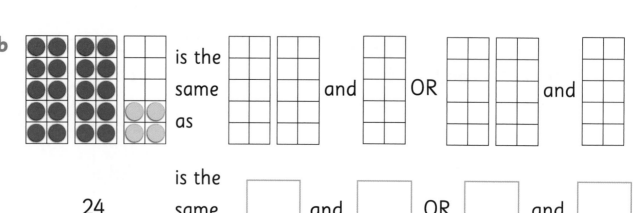

24 is the same as ☐ and ☐ OR ☐ and ☐

2 Find 4 ways to partition:

a 28 ☐ and ☐ is 28.   ☐ and ☐ is 28.

☐ and ☐ is 28.   ☐ and ☐ is 28.

b 57 ☐ and ☐ is 57.   ☐ and ☐ is 57.

☐ and ☐ is 57.   ☐ and ☐ is 57.

OXFORD UNIVERSITY PRESS

What other words do you know for take away?

15 take away **8** is **7**.

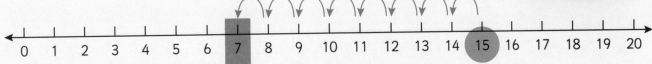

Guided practice

1 Count back to find the answers.

a 13 take away 4 is ☐ .

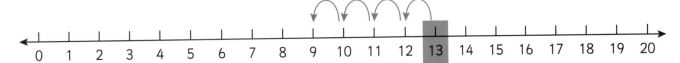

b 10 take away 7 is ☐ .

c 17 take away 5 is ☐ .

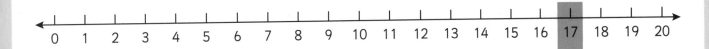

d 19 take away 6 is ☐ .

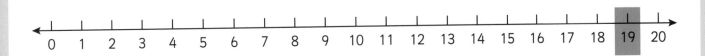

Independent practice

1 Circle the starting number. Then count back.

a　14 take away 2 is ☐ .

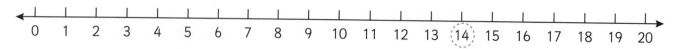

b　18 take away 8 is ☐ .

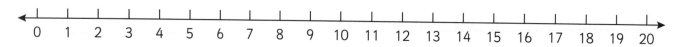

c　16 take away 12 is ☐ .

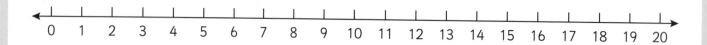

2 Show on the number line and solve.

a　20 take away 6 is ☐ .

b　20 take away 9 is ☐ .

c　19 take away 9 is ☐ .

OXFORD UNIVERSITY PRESS

3

a   How many? ☐

b   Cross out 4 and count back.

c   How many now? ☐

How could you use counting on to check your answers?

4

a   How many? ☐

b   Cross out 7 and count back.

c   How many now? ☐

5

a   Draw 23.

b   Cross out 9 and count back.

c   How many now? ☐

6

a   Draw 28.

b   Cross out 11 and count back.

c   How many now? ☐

Extended practice

1 Count back to find the answers.

a 13 take away 6 is ☐ .

b 19 take away 4 is ☐ .

c 27 take away 5 is ☐ .

d 30 take away 8 is ☐ .

2 Count back to find:

1	2	3	4	5	6	7	8	9	10
11	12	13	14	15	16	17	18	19	20
21	22	23	24	25	26	27	28	29	30
31	32	33	34	35	36	37	38	39	40
41	42	43	44	45	46	47	48	49	50
51	52	53	54	55	56	57	58	59	60
61	62	63	64	65	66	67	68	69	70
71	72	73	74	75	76	77	78	79	80
81	82	83	84	85	86	87	88	89	90
91	92	93	94	95	96	97	98	99	100

a 5 less than 37. ☐

b 7 less than 45. ☐

c 6 less than 63. ☐

d 6 less than 81. ☐

e 9 less than 36. ☐

f 8 less than 94. ☐

OXFORD UNIVERSITY PRESS

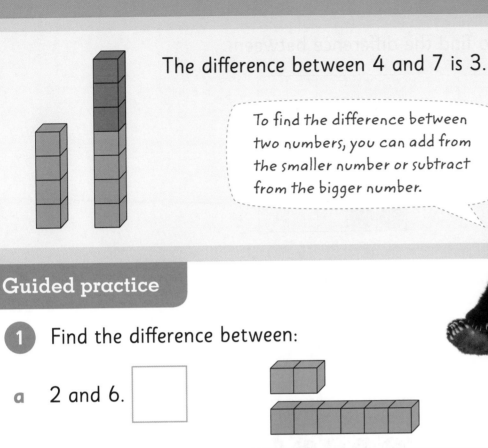

The difference between 4 and 7 is 3.

> To find the difference between two numbers, you can add from the smaller number or subtract from the bigger number.

Guided practice

1 Find the difference between:

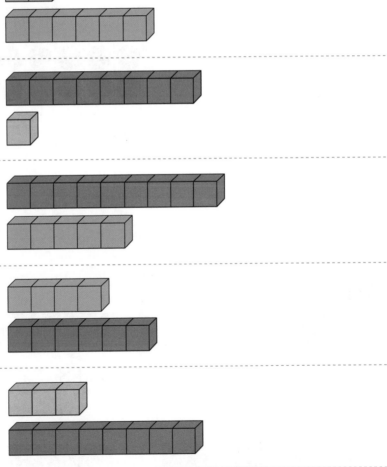

a   2 and 6.

b   8 and 1.

c   9 and 5.

d   4 and 6.

e   3 and 8.

f   11 and 8.

1 Draw more to find the difference between:

a 4 and 8.

b 3 and 9.

c 7 and 13.

d 14 and 18.

2 Circle the pairs that have a difference of 3.

1 and 4

2 and 8

6 and 3

9 and 7

7 and 4

2 and 5

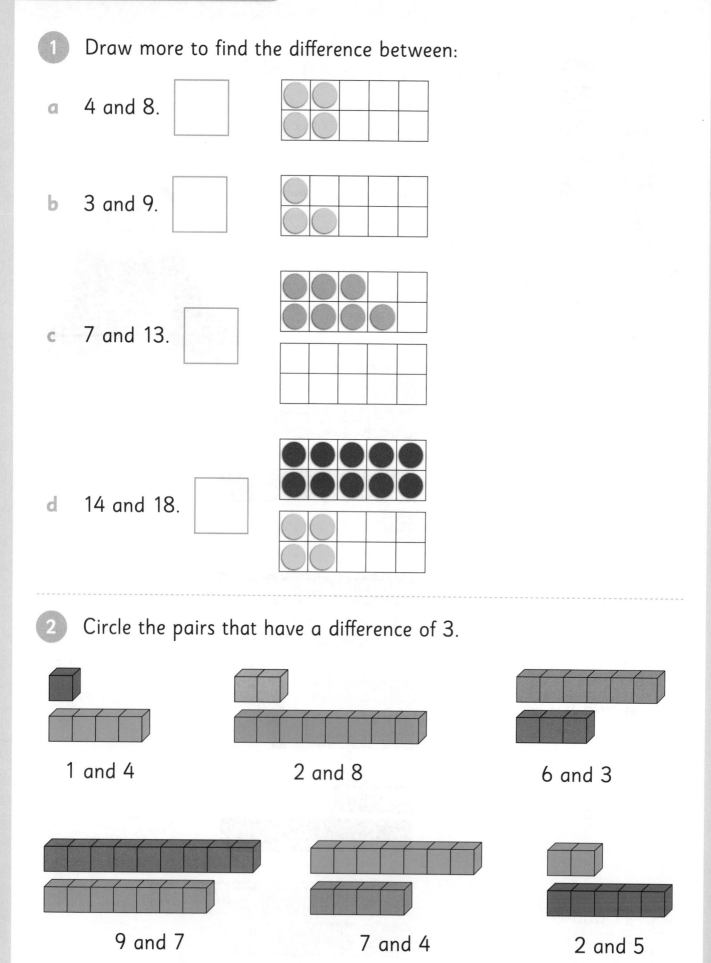

OXFORD UNIVERSITY PRESS

3 Count up to find the difference between:

a 9 and 12. ⬜

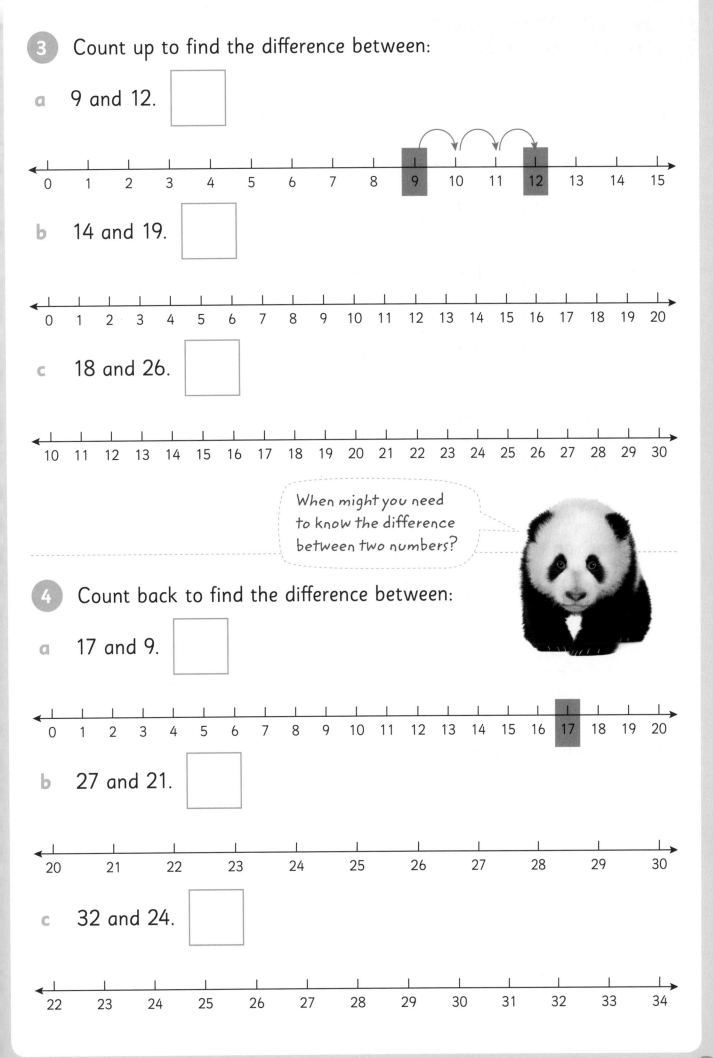

b 14 and 19. ⬜

c 18 and 26. ⬜

When might you need to know the difference between two numbers?

4 Count back to find the difference between:

a 17 and 9. ⬜

b 27 and 21. ⬜

c 32 and 24. ⬜

1 Find pairs of numbers with a difference of 4.

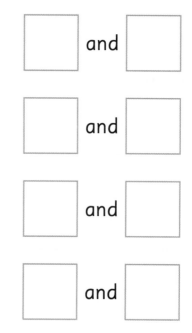

7	25	8
18	13	19
16	23	11
21	20	14

☐ and ☐

☐ and ☐

☐ and ☐

☐ and ☐

2 Show on the empty number line:

a   the difference between 25 and 29.

⟵————————————————————————⟶

b   the difference between 37 and 43.

⟵————————————————————————⟶

c   the difference between 48 and 57.

⟵————————————————————————⟶

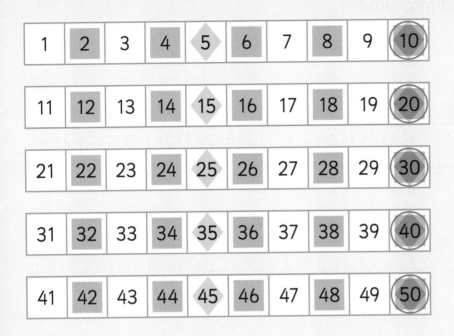

1	2	3	4	5	6	7	8	9	10
11	12	13	14	15	16	17	18	19	20
21	22	23	24	25	26	27	28	29	30
31	32	33	34	35	36	37	38	39	40
41	42	43	44	45	46	47	48	49	50

■ Counting by 2s

◆ Counting by 5s

◯ Counting by 10s

Why do you think it's called "skip counting"?

Guided practice

1 Finish by skip counting.

a Count by 2s

2	4	6	8							

b Count by 5s

5	10	15	20							

c Count by 10s

10	20	30		

1 Skip count on the number line:

a by 2.

b by 5.

c by 10.

2 Fill in the gaps.

a 2s

38	40	42		46		50			56

b 5s

35	40		50			65		75	

c 10s

10		30				70		100

3 Skip count to find how many balloons, fries, rabbits and fingers.

a

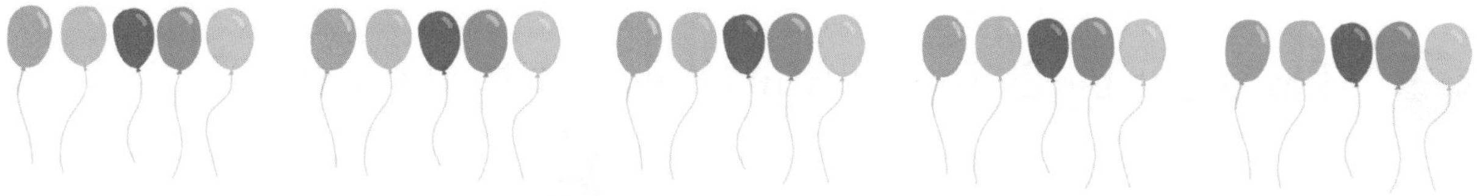

☐ ☐ ☐ ☐ ☐

b

☐ ☐ ☐ ☐ ☐

c

☐ ☐ ☐ ☐ ☐ ☐ ☐

d

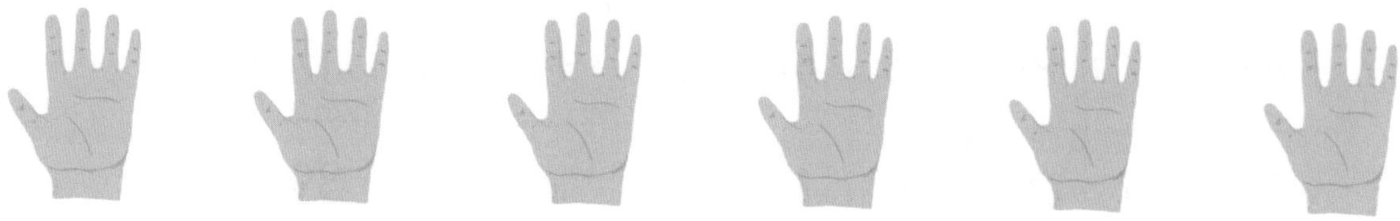

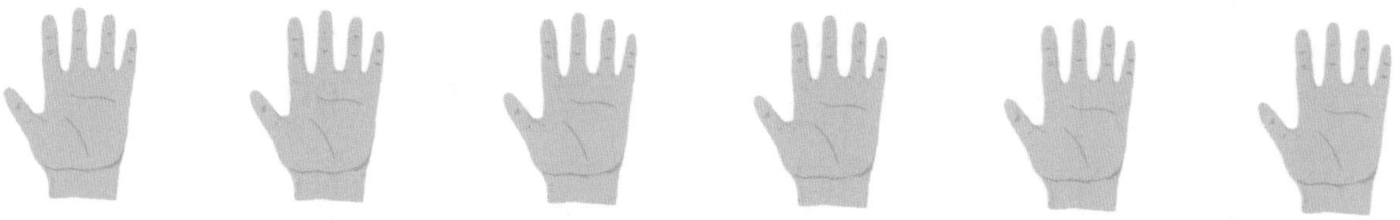

☐ ☐ ☐ ☐ ☐ ☐

☐ ☐ ☐ ☐ ☐ ☐

1 Skip count by 2s to help the koala get to the tree.

73	88	66	98	65	56	100	98
68	87	86	28	72	70	88	96
76	78	80	82	84	48	60	94
74	72	48	90	86	88	90	92
71	70	63	78	68	46	64	72

Can you see a number pattern when you skip count by 2?

- -

2 Colour the squares to skip count by 5s from 5 and find the secret number.

26	14	64	46	49	52	33	78	84	3
41	5	80	65	44	30	94	22	17	63
53	37	28	10	12	15	16	75	39	81
92	56	70	35	86	60	95	50	20	47
93	87	32	55	94	91	6	25	87	59
39	45	40	85	27	21	73	90	99	77
32	24	63	72	58	68	66	43	51	31

Secret number: [   ]

OXFORD UNIVERSITY PRESS

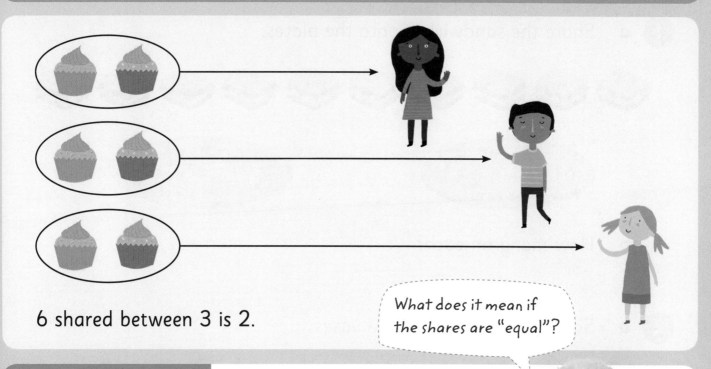

6 shared between 3 is 2.

What does it mean if the shares are "equal"?

Guided practice

1 Complete the sentences.

a

8 shared between 2 is ☐.

b

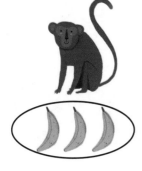  

9 shared between 3 is ☐.

1 a Share the sandwiches onto the plates.

b How many on each? [ ]

2 a Share the flowers into the vases.

b How many in each? [ ]

3 a Share the mushrooms onto the pizzas.

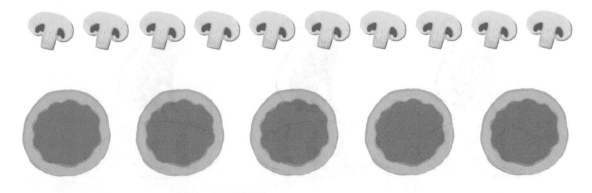

b How many on each? [ ]

OXFORD UNIVERSITY PRESS

4 Complete the number sentences.

a

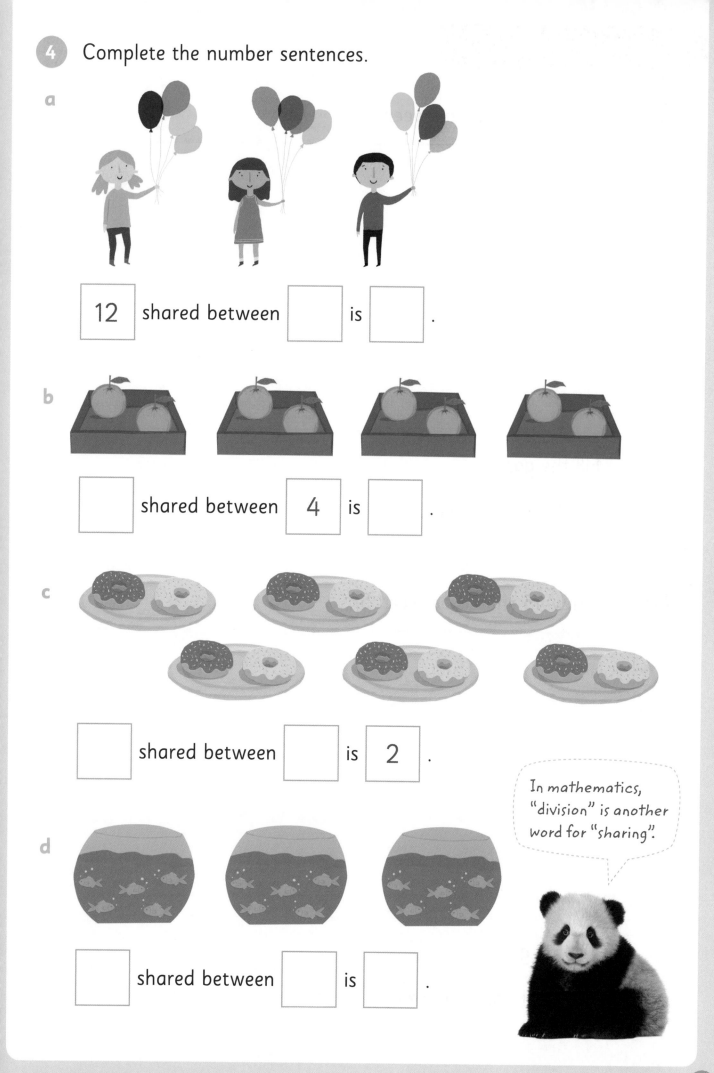

12 shared between ☐ is ☐ .

b

☐ shared between 4 is ☐ .

c

☐ shared between ☐ is 2 .

In mathematics, "division" is another word for "sharing".

d

☐ shared between ☐ is ☐ .

1 **a** Draw 12 shared between 3.

b Fill the gap.

12 shared between 3 is ☐ .

2 **a** Draw 15 shared between 5.

b Fill the gaps.

☐ shared between ☐ is ☐ .

A **cardinal** number tells you how many things there are. An **ordinal** number shows the order or position of something.

1	2	3	4	5	6
one	two	three	four	five	six
1st	2nd	3rd	4th	5th	6th
first	second	third	fourth	fifth	sixth

Guided practice

1 Follow the instructions to colour the mice.

| 1 | 2 | 3 | 4 | 5 | 6 |

a 1st: **red**   b 2nd: **grey**   c 3rd: **purple**

d 4th: **blue**   e 5th: **yellow**   f 6th: **green**

2 What colour is:

a the 1st?

b the 2nd?

c the 6th?

1 Match the words and numbers.

1	2	3	4	5	6

three	six	one	five	four	two

2 Match the words and numbers.

first	second	third	fourth	fifth	sixth

3rd	1st	6th	5th	2nd	4th

3 Label the dogs from 1st to 6th.

4 Rewrite in the correct order.

second	fourth	third	first

OXFORD UNIVERSITY PRESS

5 Look at the picture.

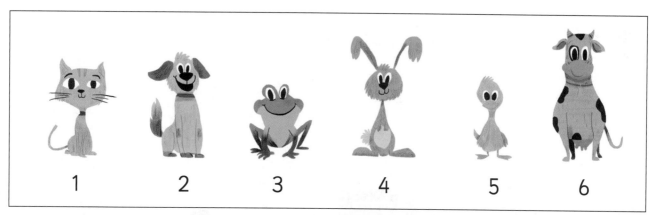

| 1 | 2 | 3 | 4 | 5 | 6 |

Which animal is:

a   1st?

b   6th?

c   second?

d   third?

What comes after 6th?

- -

6 Circle the:

a   2nd.

b   5th.

c   4th.

Extended practice

1 Match the activities to their order.

| step 3 | step 2 | step 4 | step 1 |

| 4th | 1st | 3rd | 2nd |

2 Number each box. Then draw a:

a ☐ in the 1st box.

b ◯ in the third box.

c △ in the 6th box.

d ☺ in the last box.

e 🌳 in the 4th box.

f 🧍 in the second box.

OXFORD UNIVERSITY PRESS

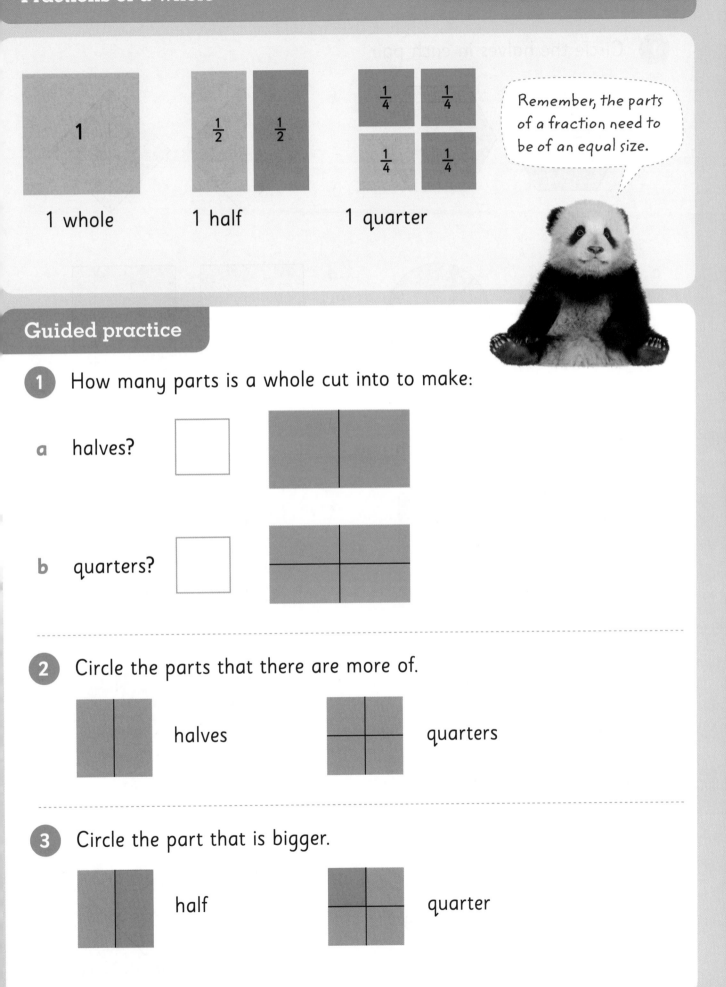

1 whole   1 half   1 quarter

Remember, the parts of a fraction need to be of an equal size.

Guided practice

1. How many parts is a whole cut into to make:

a. halves?

b. quarters?

2. Circle the parts that there are more of.

halves   quarters

3. Circle the part that is bigger.

half   quarter

1 Circle the halves in each pair.

a 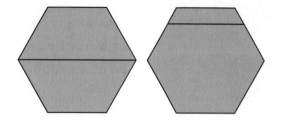   b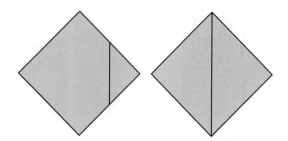

c 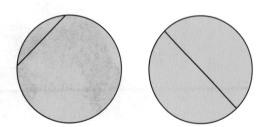   d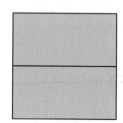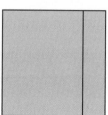

2 Circle the quarters in each pair.

a 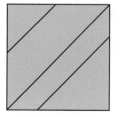   b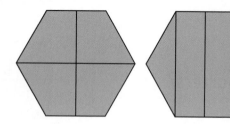

c 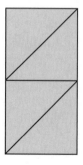 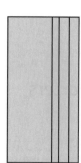   d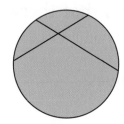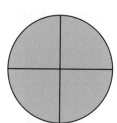

e 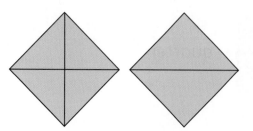   f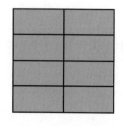

3 Draw lines to make halves.

a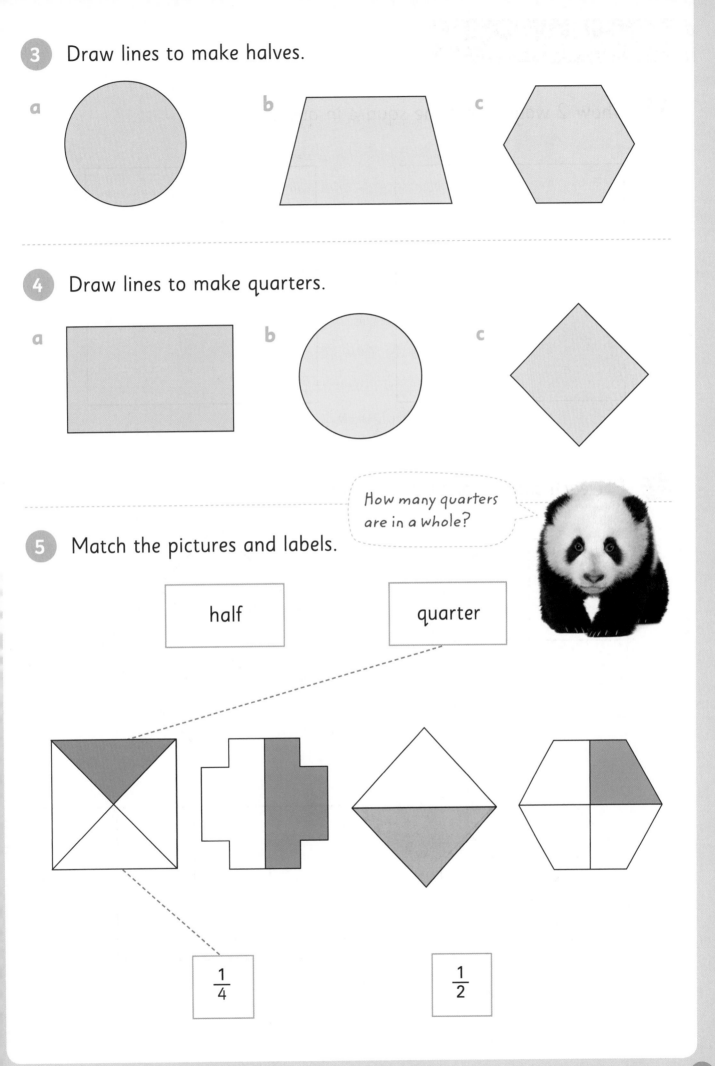

b

c

4 Draw lines to make quarters.

a

b

c

How many quarters
are in a whole?

5 Match the pictures and labels.

half

quarter

$\frac{1}{4}$

$\frac{1}{2}$

1 Show 2 ways to cut the square in quarters.

---

2 For each shape, colour $\frac{1}{2}$ blue and $\frac{1}{4}$ red.

a

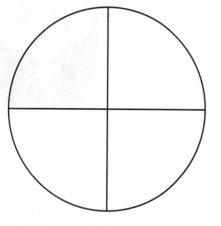

b

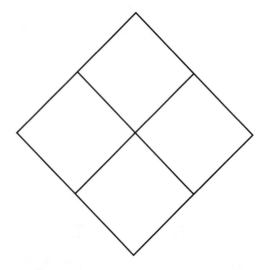

c

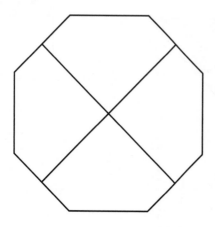

d

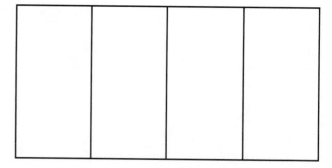

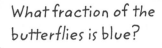

What fraction of the butterflies is blue?

There are eight butterflies.

Half are green.

One quarter are red.

Guided practice

1 Halves or quarters?

a

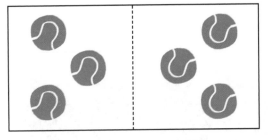

halves	quarters

b

halves	quarters

c

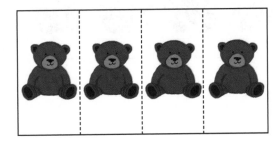

halves	quarters

d

halves	quarters

1

a   Draw circles to divide the group into halves.

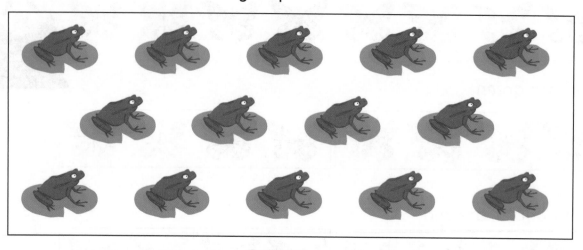

b   How many groups?

c   How many in each group?

2

a   Draw circles to divide the group into quarters.

b   How many groups?

c   How many in each group?

OXFORD UNIVERSITY PRESS

3 Draw more to make equal halves.

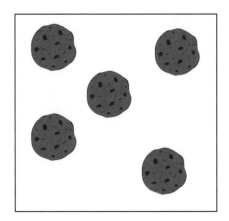
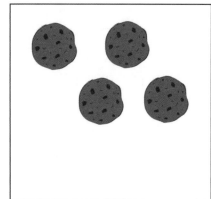

$\frac{1}{2}$ means one part out of two.

4 Draw more to make equal quarters.

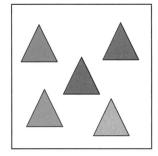
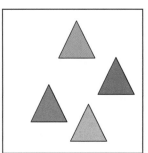
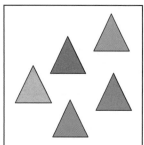
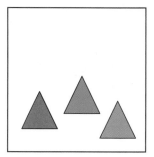

5 Match the words, pictures and symbols.

halves

quarters

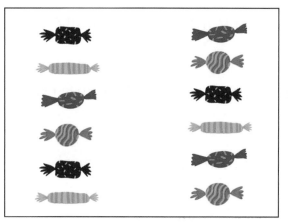

$\frac{1}{4}$

$\frac{1}{2}$

1

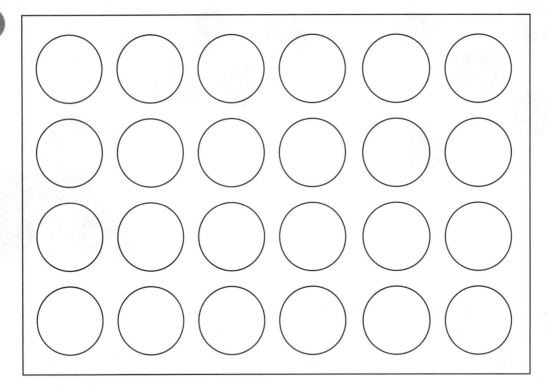

a   How many circles?

b   Colour half **red**.

c   Colour one quarter **blue**.

d   How many in one half?

e   How many in one quarter?

f   Which group is bigger?

half	quarter

g   Which fraction has more groups?

halves	quarters

h   What fraction is left uncoloured?

half	quarter

The size of coins does not relate to their value.

 This 50c coin is big in size BUT

 this $2 coin has a greater value.

The biggest coin doesn't always have the greatest value!

Guided practice

1 Draw a line to match each coin with its value.

| 5 cents | 10 cents | 20 cents | 50 cents | 1 dollar | 2 dollars |

2

a List the coins in order of **value** from most to least.

b Which coins are worth more than 50c?

c Which coin is worth the least?

a Number the coins in order of **size** from smallest to biggest.

b Which coins are bigger than a $1 coin?

c Which gold coin is the smallest in size?

2 Circle the coin that is worth the most in each group.

a

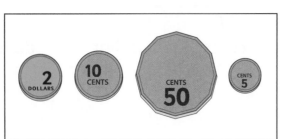

b

3 Circle the coin that is worth the least in each group.

a

b

4 Draw the coins in order of **value** from least to most.

a
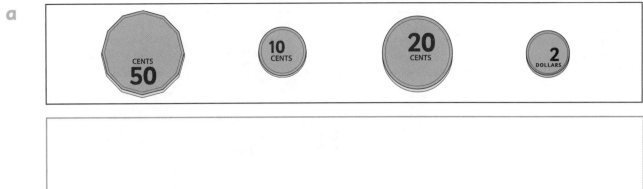

b

5 Draw in order of **size** from smallest to biggest.

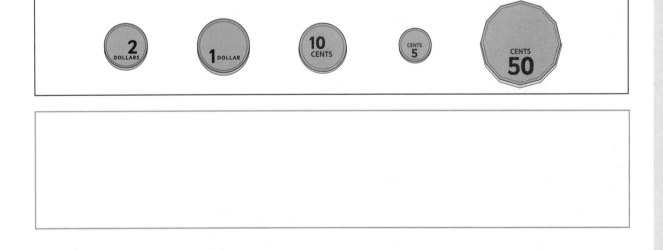

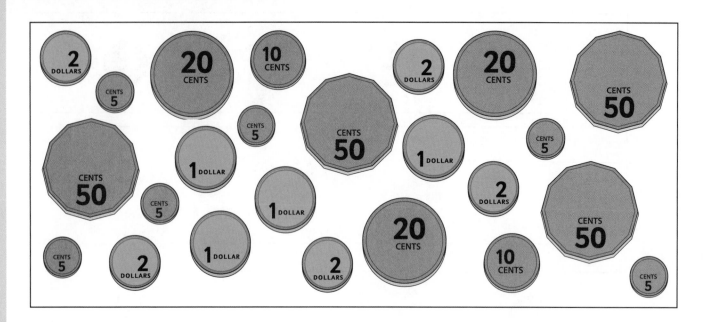

1 a How many 5c coins?

   b What is their total value?

2 a How many 50c coins?

   b What is their total value?

3 a How many $2 coins?

   b What is their total value?

4 What is the total value of:

   a 20 CENTS and 10 CENTS

   b 1 DOLLAR and 2 DOLLARS

OXFORD UNIVERSITY PRESS

The rule for this colour pattern is red, green.

The rule for this letter pattern is A, B.

What other sorts of patterns are there?

Guided practice

1 Finish the colour patterns.

a

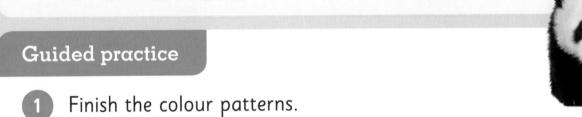

b

2 Finish the letter patterns.

a

| Y | Z | Y | Z | Y | | |

b

| A | B | C | A | B | | |

Independent practice

This is a 2 pattern. 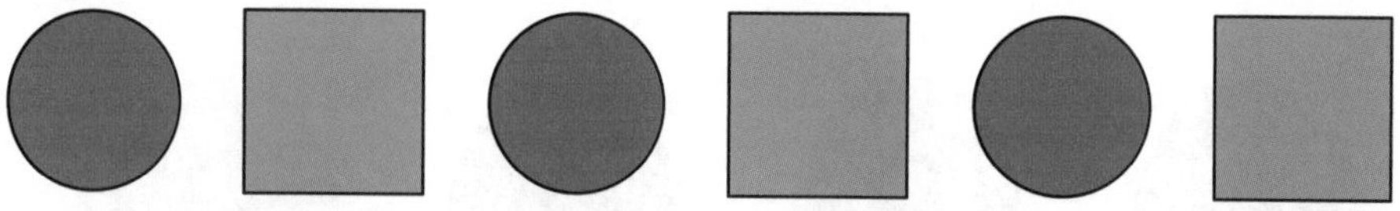

This is a 3 pattern. 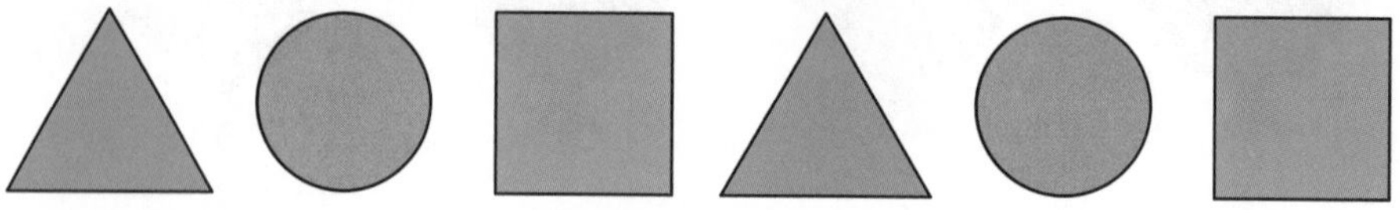

1 Are these 2 or 3 patterns?

a

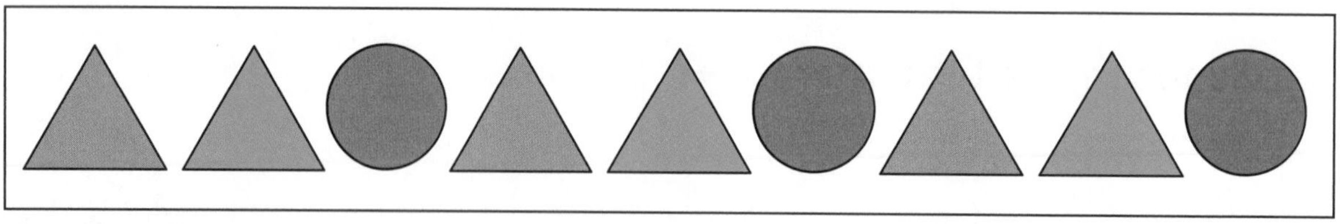

2	3

b

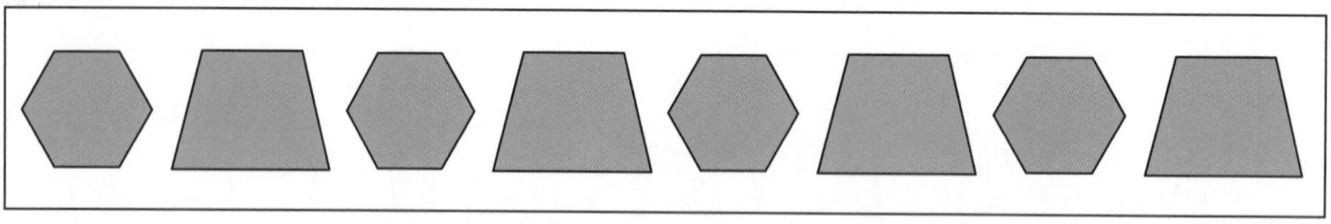

2	3

c

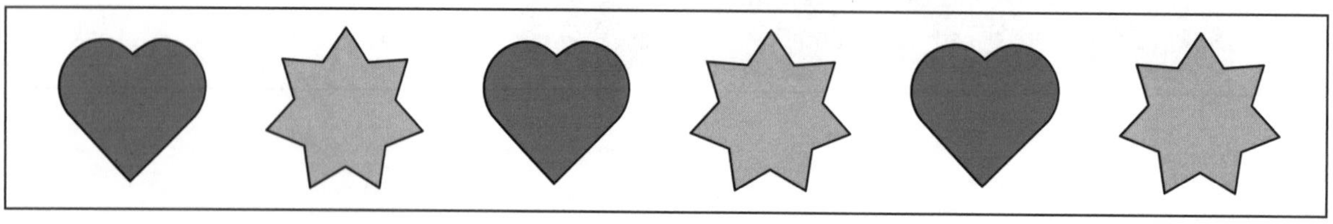

2	3

d

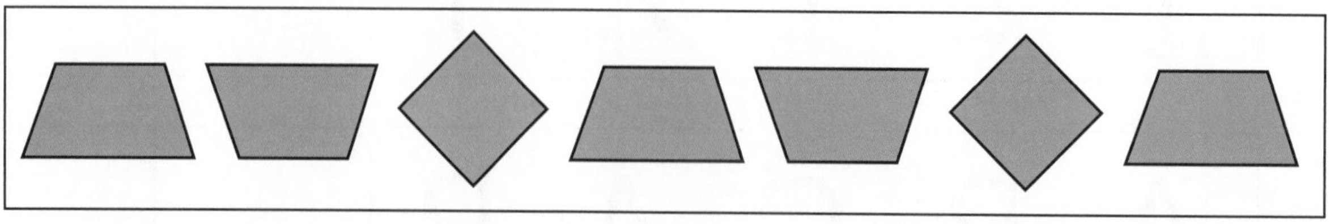

2	3

2 **a** Continue the pattern.

b What is the rule?

> This type of pattern is called a repeating pattern.

3 **a** Continue the pattern.

b What is the rule?

4 **a** Circle the error.

b What should the colour be?

5 **a** Circle the error.

B B A B B B A A B A

b What should the letter be?

1 Finish the growing patterns.

a

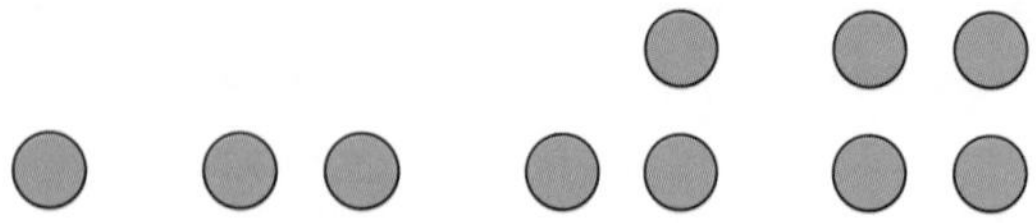

b

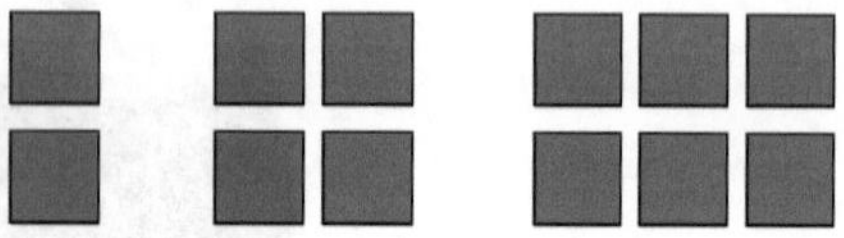

2

a Create a colour pattern.

b What is the rule?

3

a Create a shape pattern.

b What is the rule?

OXFORD UNIVERSITY PRESS

Counting by 2

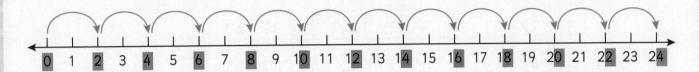

Each number has its own counting pattern.

Guided practice

1

a Circle the numbers in the 2s counting pattern.

1	2	3	4	5	6	7	8	9	10
11	12	13	14	15	16	17	18	19	20
21	22	23	24	25	26	27	28	29	30

b Which 5 digits repeat?

c Count on by 2s.

32

1 **a** Circle the numbers in the 5s counting pattern.

1	2	3	4	5	6	7	8	9	10
11	12	13	14	15	16	17	18	19	20
21	22	23	24	25	26	27	28	29	30
31	32	33	34	35	36	37	38	39	40
41	42	43	44	45	46	47	48	49	50

b Which 2 digits repeat? ☐ ☐

c Count on by 5s. | 55 | ☐ ☐ ☐

2 **a** Circle the numbers in the 10s counting pattern.

1	2	3	4	5	6	7	8	9	10
11	12	13	14	15	16	17	18	19	20
21	22	23	24	25	26	27	28	29	30
31	32	33	34	35	36	37	38	39	40
41	42	43	44	45	46	47	48	49	50
51	52	53	54	55	56	57	58	59	60
61	62	63	64	65	66	67	68	69	70
71	72	73	74	75	76	77	78	79	80
81	82	83	84	85	86	87	88	89	90
91	92	93	94	95	96	97	98	99	100

b Which digit repeats? ☐

c Are the numbers odd or even? | odd | even |

OXFORD UNIVERSITY PRESS

3

1	2	3	4	5	6	7	8	9	10
11	12	13	14	15	16	17	18	19	20
21	22	23	24	25	26	27	28	29	30
31	32	33	34	35	36	37	38	39	40
41	42	43	44	45	46	47	48	49	50
51	52	53	54	55	56	57	58	59	60
61	62	63	64	65	66	67	68	69	70
71	72	73	74	75	76	77	78	79	80
81	82	83	84	85	86	87	88	89	90
91	92	93	94	95	96	97	98	99	100

a Circle all the numbers that have the digit 4 in them.

b How many?

c Colour the numbers with the digit 9.

d How many?

- -

4 Fill in the gaps.

a

35	40	45				65			80

Counting by?

2	5	10

b

40	50		70			100

Counting by?

2	5	10

> How do the counting patterns help you to know what number comes next?

c

20	22	24		28				36	

Counting by?

2	5	10

1	2	3	4	5	6	7	8	9	10
11	12	13	14	15	16	17	18	19	20
21	22	23	24	25	26	27	28	29	30
31	32	33	34	35	36	37	38	39	40
41	42	43	44	45	46	47	48	49	50
51	52	53	54	55	56	57	58	59	60
61	62	63	64	65	66	67	68	69	70
71	72	73	74	75	76	77	78	79	80
81	82	83	84	85	86	87	88	89	90
91	92	93	94	95	96	97	98	99	100

1　**a** Circle 28 in blue.　　**b** Circle 10 more than 28.

　　c Circle 10 less than 28.　**d** Which digit repeats?

2　**a** Circle 54 in red.　　**b** Circle 5 more than 54.

　　c Circle 5 less than 54.

　　d What number would come next in the pattern?

3 Finish the patterns.

a

100	102			108	110				

b

105	110	115		125			140		

c

100		120	130			160			

Length

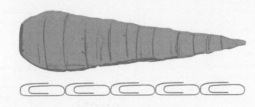

The carrot is
5 paperclips long.

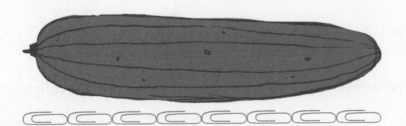

The cucumber is
8 paperclips long.

Guided practice

Make sure you don't leave any gaps between the paperclips when you are measuring length.

1 How long?

a

☐ paperclips long

b

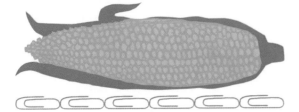

☐ paperclips long

c

☐ paperclips long

2 Which is the longest?

corn 	zucchini 	chilli 

1 Estimate and measure with paperclips:

a the length of your pencil.

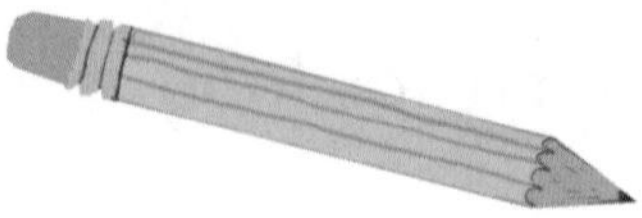

estimate: ☐ paperclips

length: ☐ paperclips

b the length of your student book.

estimate: ☐ paperclips

length: ☐ paperclips

c the length of your shoe.

estimate: ☐ paperclips

length: ☐ paperclips

d the length of this line.

estimate: ☐ paperclips

length: ☐ paperclips

2 Draw the **shortest** item from question 1.

Area measures the surface of something.

I wonder what my area is.

The placemat has an area of 12 tiles.

The photo has an area of 6 tiles.

Guided practice

1 Find the area.

a  [ ] tiles

b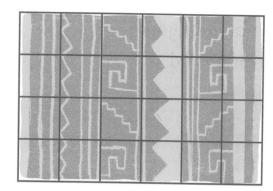

[ ] tiles

c   [ ] tiles

d Which has the **smallest** area?

book	rug 	calendar

1 Estimate and measure with tiles or blocks:

a the area of your student book.
estimate: ☐ tiles or blocks

area: ☐ tiles or blocks

b the area of a poster.
estimate: ☐ tiles or blocks

area: ☐ tiles or blocks

c the area of your lunch box lid.
estimate: ☐ tiles or blocks

area: ☐ tiles or blocks

d the area of this rectangle.
estimate: ☐ tiles or blocks

area: ☐ tiles or blocks

2 Which has the **biggest** area? ☐

OXFORD UNIVERSITY PRESS

Extended practice

Length

1 Find the length of your desk:

a using pencils to measure.
[ ] pencils

b using pencil cases to measure.
[ ] pencil cases

2 Which did you need more of?

pencils	pencil cases

Area

3 Find the area of this book:

a using blocks to measure.
[ ] blocks

b using sticky notes to measure.
[ ] sticky notes

4 Which did you need more of?

blocks	sticky notes

5 Find an object with a smaller area than this book.

a Measure the area of your object using blocks.
[ ] blocks

b Measure the area of your object using sticky notes.
[ ] sticky notes

6 Which did you need more of?

blocks	sticky notes

Volume is how much space an object takes up.

This box has a
volume of 4 cubes.

This box has a
volume of 6 cubes.

Which of the two
boxes has the
bigger volume?

Guided practice

1 Write the volume of these objects.

a

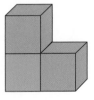

☐ cubes

b

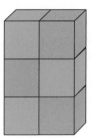

☐ cubes

c

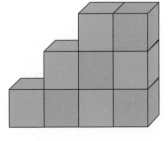

☐ cubes

d

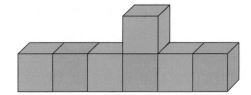

☐ cubes

OXFORD UNIVERSITY PRESS

Independent practice

1 Use cubes to make each object. Record the volume.

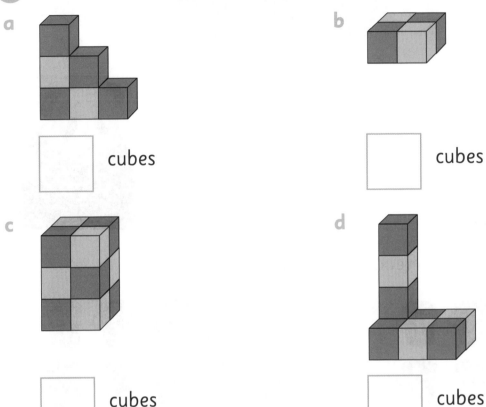

a ☐ cubes

b ☐ cubes

c ☐ cubes

d ☐ cubes

2 a Circle in **blue** the object that needed the **most** cubes.

b Circle in **red** the object that needed the **fewest** cubes.

3

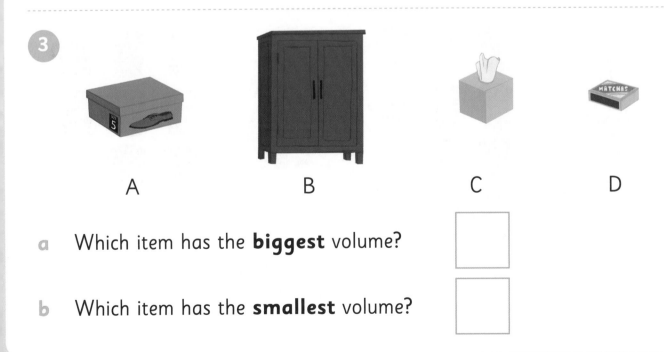

A   B   C   D

a Which item has the **biggest** volume? ☐

b Which item has the **smallest** volume? ☐

Capacity is how much a container can hold.

Which of the two bowls has the bigger capacity?

This bowl has a capacity of 4 cups.

This bowl has a capacity of 10 cups.

Guided practice

1 Write the capacity of each jug in cups.

a

[ ] cups

b

[ ] cups

c

[ ] cups

d

[ ] cups

OXFORD UNIVERSITY PRESS

1 Circle the unit you would use to measure the capacity of the items.

a

b

c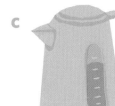

d

2

a Draw an item with a **bigger** capacity.

b Draw an item with a **smaller** capacity.

c Circle the unit you would use to measure the capacity of the items you drew.

1

a Make and draw an object with a volume of 8 cubes.

b Make and draw a different object with a volume of 8 cubes.

2 Find a cup and two larger containers.

a Draw your containers.

b Estimate the capacity of each container.

☐ cups   ☐ cups

c Measure and record the capacities.

☐ cups   ☐ cups

Guided practice

1 Draw something **lighter** on the scales.

a

b

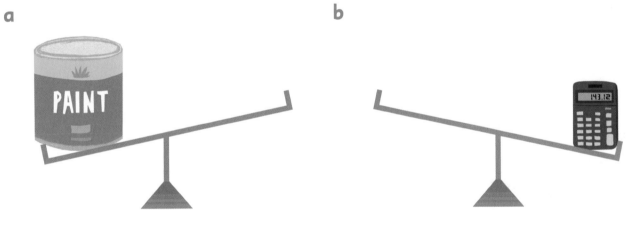

2 Draw something **heavier** on the scales.

a

b

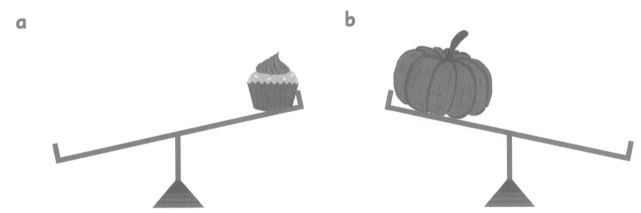

1 Choose pairs of items from below and draw them on the correct sides of the scales.

a   b

c   d

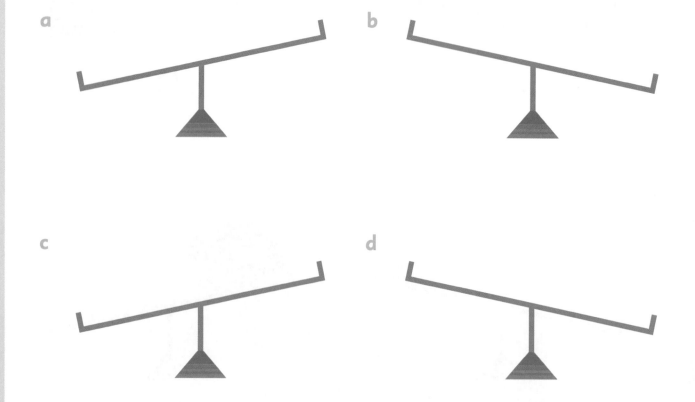

OXFORD UNIVERSITY PRESS

2 Estimate and then check if each item is lighter or heavier than your pencil case.

a a ruler

My estimate:

lighter	heavier

Result:

lighter	heavier

b a stapler

My estimate:

lighter	heavier

Result:

lighter	heavier

c this book

My estimate:

lighter	heavier

Result:

lighter	heavier

d a pencil

My estimate:

lighter	heavier

Result:

lighter	heavier

e a drink bottle

My estimate:

lighter	heavier

Result:

lighter	heavier

f scissors

My estimate:

lighter	heavier

Result:

lighter	heavier

How will you know which items are lighter than your pencil case?

1 Collect some counters and cubes.

Use a scale to find how many counters balance:

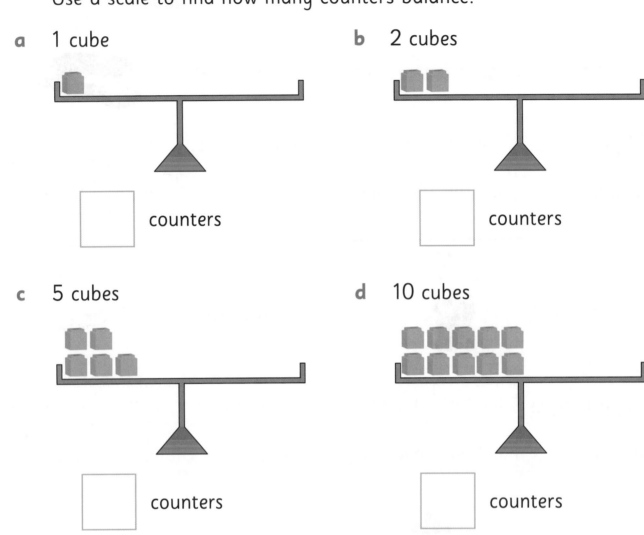

a 1 cube

☐ counters

b 2 cubes

☐ counters

c 5 cubes

☐ counters

d 10 cubes

☐ counters

2 Redraw the items from **lightest** to **heaviest**.

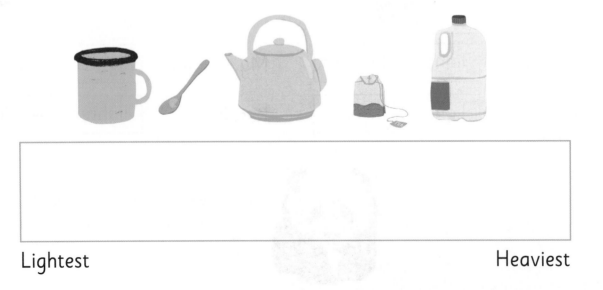

Lightest	Heaviest

OXFORD UNIVERSITY PRESS

1 o'clock

half past 1

2 o'clock

half past 2

Guided practice

Where is the hour hand for o'clock time? Where is it for half past time?

1 O'clock or half past?

a

| o'clock | half past |

b

| o'clock | half past |

c

| o'clock | half past |

d

| o'clock | half past |

e

| o'clock | half past |

f

| o'clock | half past |

2 What is the time?

a

☐ o'clock

b

half past

c

half past

Independent practice

1 Match the o'clock times.

| 7 o'clock | 11 o'clock | 2 o'clock | 6 o'clock |

2 Show:

a

1 o'clock

b

8 o'clock

c

3 o'clock

d

9 o'clock

e

12 o'clock

f

4 o'clock

3 Match the half past times.

| half past 3 | half past 11 | half past 7 | half past 2 |

4 Show:

a

half past 8

b

half past 1

c

half past 9

d

half past 12

e

half past 6

f

half past 5

Extended practice

1 Match the clocks and their times.

6:30	12:00	5:30	12:30	5:00

2 Write the times.

a

     :

b

     :

c

     :

d

     :

e

     :

f

     :

3 Write the time in words and numbers.

Words:

Numbers: :

OXFORD UNIVERSITY PRESS

A car trip

HOME   BEACH

2 hours

The weekend

2 days

A holiday

2 weeks

Winter

3 months

"Duration" means how long something lasts.

Guided practice

1 Hours, days, weeks or months?

a   12 [   ] in a year.

b   24 [   ] in a day.

c   30 [   ] in April.

d   4 [   ] in February.

e $6\frac{1}{2}$ [   ] in a school day.

f   3 [   ] in spring.

Independent practice

1 Draw an event that takes **longer than**:

a

4 hours.

b

1 day.

2

a Match the events and durations.

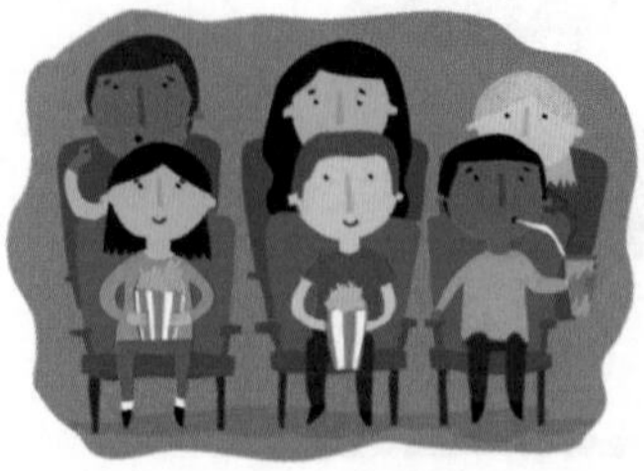

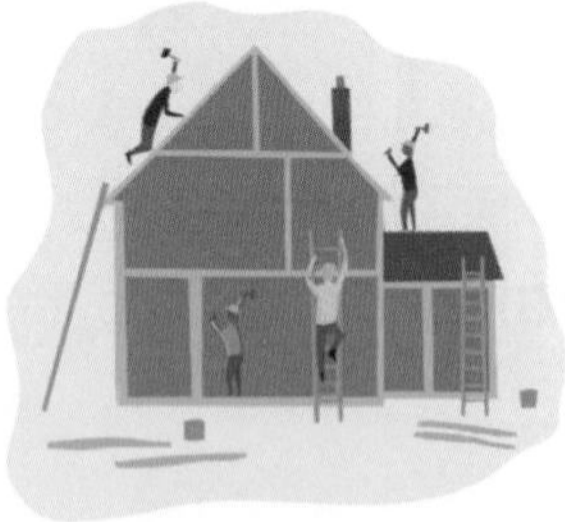

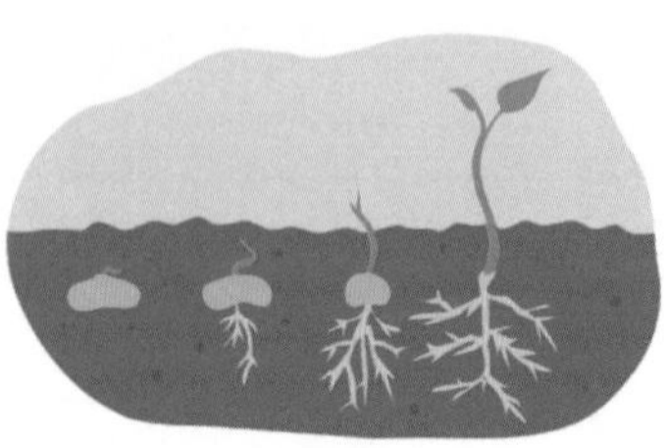

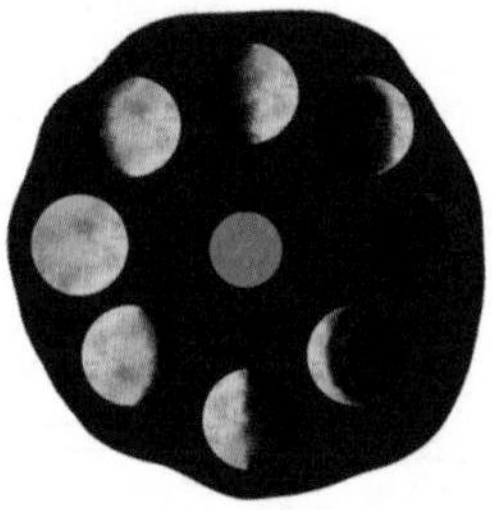

5 days	4 weeks	2 hours	4 months

b Which event is the shortest?

3 Number from **shortest** to **longest** duration.

a

the weekend

☐

caterpillar to butterfly

☐

birthday party

☐

b

sleeping at night

☐

the school week

☐

growing long hair

☐

c

time until your birthday

☐

reading a chapter book

☐

a netball game

☐

4 Draw the event from question 3 that lasts the **longest**.

Would you measure it in hours, days, months or years?

Extended practice

1 How many:

a hours in a day? ☐

b days in a week? ☐

c weeks in a month? ☐

d months in a year? ☐

2 How long until:

a your birthday?

☐

b the end of the school day?

☐

c the end of the month?

☐

d the end of the school term?

☐

e dinner time?

☐

f the weekend?

☐

3 Draw something that takes:

a a long time.

☐

How long? ☐

b a short time.

☐

How long? ☐

OXFORD UNIVERSITY PRESS

This rectangle has:

2 **horizontal** lines   4 corners

2 **vertical** lines   4 sides

Which way do horizontal lines go? Which way do vertical lines go?

Guided practice

1   How many:

a

horizontal lines? ☐   corners? ☐

vertical lines? ☐   sides? ☐

b

horizontal lines? ☐   corners? ☐

vertical lines? ☐   sides? ☐

c

horizontal lines? ☐   corners? ☐

vertical lines? ☐   sides? ☐

d

horizontal lines? ☐   corners? ☐

vertical lines? ☐   sides? ☐

1 Colour the shapes with:

a 1 horizontal line in **green**.

b 2 vertical lines in **red**.

All four-sided shapes are **quadrilaterals**. How many quadrilaterals can you see? What other names do they have?

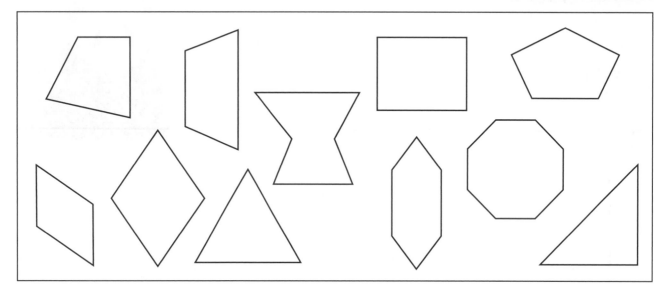

- -

2 Match the shapes and descriptions.

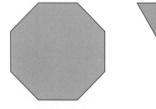  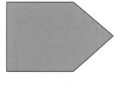  

1 vertical side	2 vertical sides	2 vertical sides	0 vertical sides	0 vertical sides
2 horizontal sides	0 horizontal sides	2 horizontal sides	1 horizontal side	0 horizontal sides
5 sides in total	6 sides in total	8 sides in total	3 sides in total	4 sides in total
5 corners	6 corners	8 corners	3 corners	4 corners
Pentagon	**Hexagon**	**Octagon**	**Triangle**	**Quadrilateral**

3 Colour the:

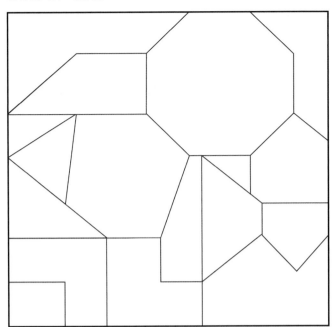

a triangles blue.

b quadrilaterals red.

c pentagons yellow.

d hexagons green.

e octagons purple.

> Parallel lines are two or more lines that are the same distance apart and never cross.

4 Parallel or not parallel?

a

parallel	not parallel

b

parallel	not parallel

c

parallel	not parallel

d

parallel	not parallel

e

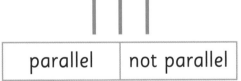

parallel	not parallel

f

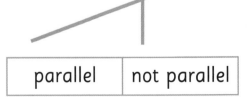

parallel	not parallel

Extended practice

1 Draw:

a a quadrilateral with 2 horizontal sides.

b a triangle with 1 vertical side.

- -

2 Name and describe.

a

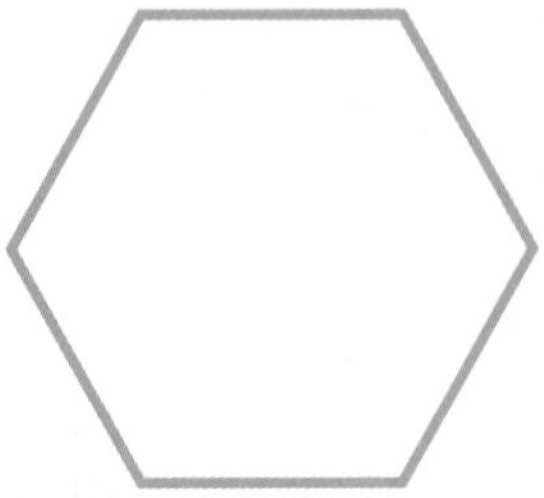

b

The faces of a rectangular prism are rectangles.

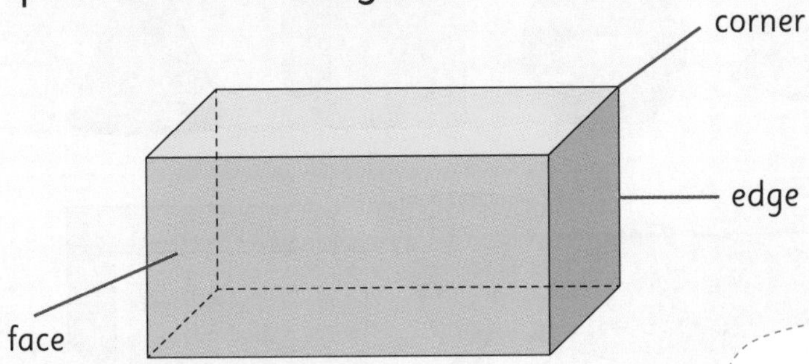

corner

edge

face

A face is a flat surface of a 3D shape.

Guided practice

1 Match the labels to the picture.

face

corner

edge

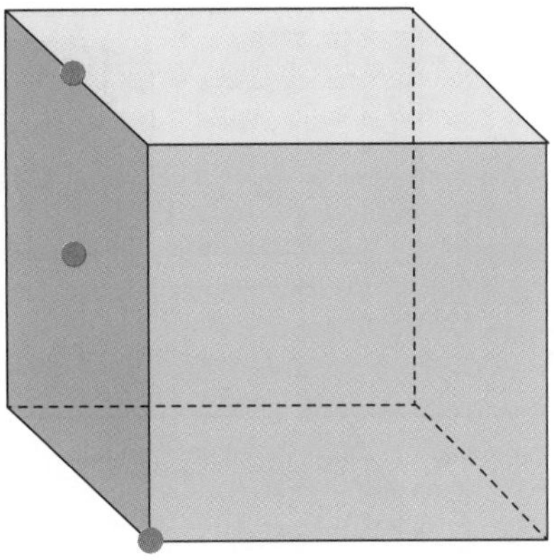

---

2 Tick the 3D shapes with a circle-shaped face.

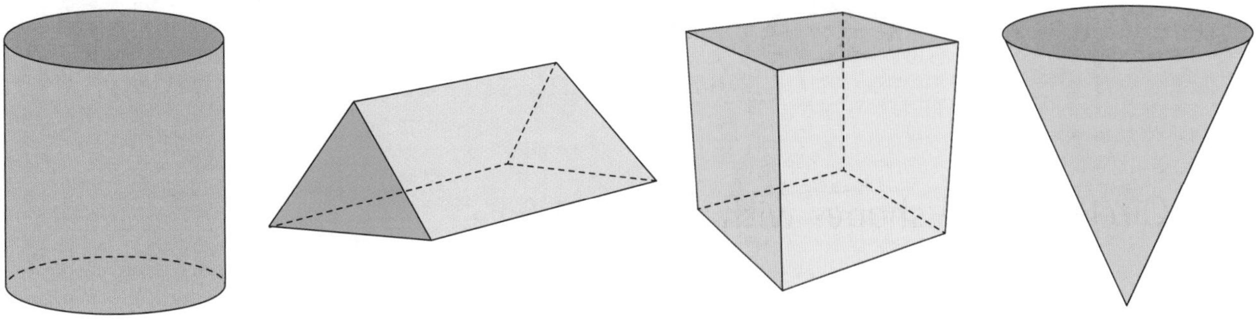

cylinder

triangular prism

cube

cone

1 Circle the 3D shape with:

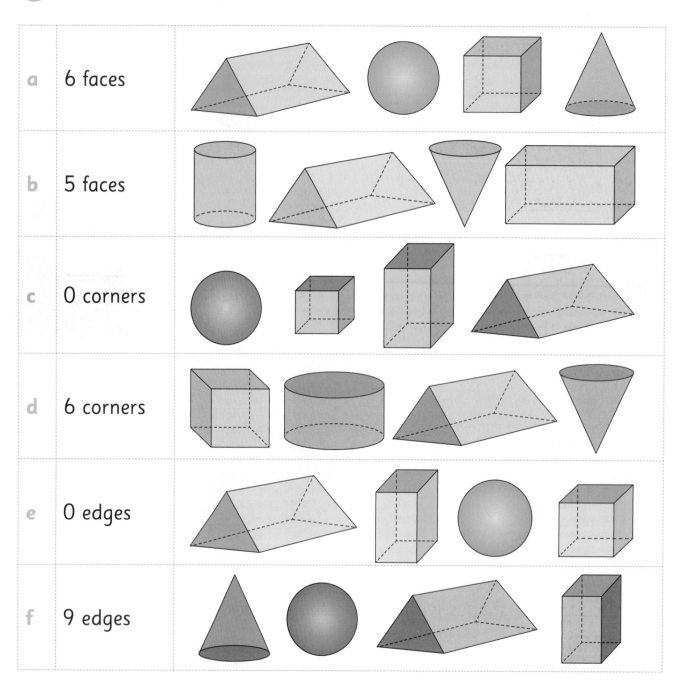

a	6 faces	
b	5 faces	
c	0 corners	
d	6 corners	
e	0 edges	
f	9 edges	

2 Circle the 3D shapes with a curved face.

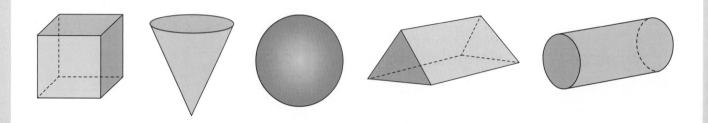

OXFORD UNIVERSITY PRESS

3 How many:

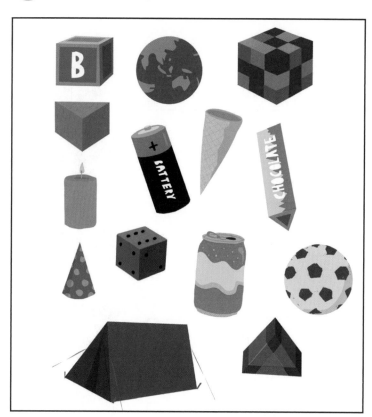

a  cubes? ☐

b cones? ☐

c triangular prisms? ☐

d spheres? ☐

e cylinders? ☐

A prism has two ends that are the same shape. All the other faces are rectangles.

4 Match the 3D shapes and faces.

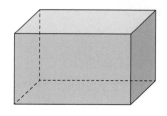 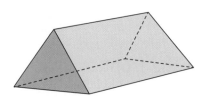 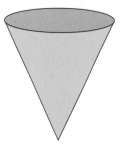 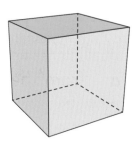

 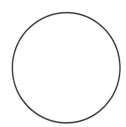  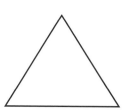

1 Who am I?

a I have:
- 6 faces
- 8 corners
- 12 edges.

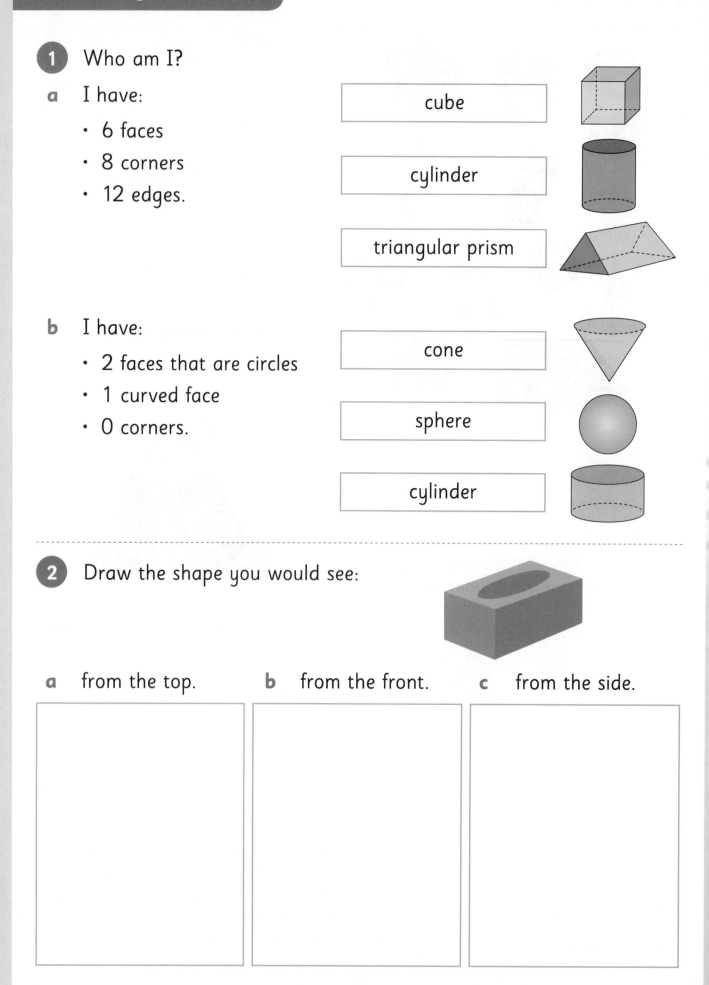

cube

cylinder

triangular prism

b I have:
- 2 faces that are circles
- 1 curved face
- 0 corners.

cone

sphere

cylinder

2 Draw the shape you would see:

a from the top.

b from the front.

c from the side.

The cat is **on** the chair.

The mouse is **under** the chair.

The dog is **in** the box.

What other words can you use to describe the position of something?

Guided practice

1 Where is:

a the bird?

on the bench	in the tree	under the car

b the car?

in the shed	in the tree	on the bench

c the cat?

on the car	in the shed	next to the tree

d the snake?

on the bench	in the tree	under the car

1 In the box below, draw:

a a cat **under** the table.

b a ball **on** the rug.

How would you describe where I am sitting on the page?

c a chair **next to** the ball.

d a book **in** the bookshelf.

e a person **between** the table and the bookshelf.

OXFORD UNIVERSITY PRESS

How many different ways can you describe where the train is?

2 What is:

a   next to the brown bear?

b   under the robot?

c   between the boat and the drum?

d   above the ball?

3 Where is:

a   the panda?

b   the drum?

1 Describe the position of:

a the pirate.

b the treasure.

2 a Draw a dog on the map.

b Describe where you drew it.

clockwise

anticlockwise

forwards

backwards

Clockwise is the direction the hands on a clock move. Anticlockwise is the opposite direction.

Guided practice

1 Clockwise or anticlockwise?

a

| clockwise | anticlockwise |

b

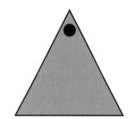

| clockwise | anticlockwise |

c

| clockwise | anticlockwise |

d

| clockwise | anticlockwise |

2 Forwards or backwards?

a

| forwards | backwards |

b

| forwards | backwards |

1 Which way should the cat move to:

a find the mouse first?

clockwise	anticlockwise

b find the fish first?

clockwise	anticlockwise

2 Which way should the hippo move to:

a find the lion first?

clockwise	anticlockwise

b find the zebra first?

clockwise	anticlockwise

3 Which way should the giraffe move to:

a find the lion first?

clockwise	anticlockwise

b find the zebra first?

clockwise	anticlockwise

How would you describe where the lion is?

OXFORD UNIVERSITY PRESS

START

4 Colour the path as you go.

a Move forward 4 spaces from START.

b Turn to the right.

c Move forward 3 spaces.

d Turn to the left.

e Move forward 2 spaces.

f Turn to the right.

g Move forward 2 spaces.

h Where are you?

> Is a right turn clockwise or anticlockwise?

1 Write directions to get from:

a the sandpit to the see-saw.

b the swings to the slide.

Animals in the park

How many dogs are there? How many rabbits?

	Dogs	Ducks	Rabbits
5	✓		
4	✓		
3	✓	✓	
2	✓	✓	
1	✓	✓	✓

Guided practice

1 Use ticks to show how many animals are on the farm.

	Goats	Horses	Chickens
5			
4			
3			
2			
1			

1 How many:

a triangles? ☐   b rectangles? ☐

c circles? ☐   d hexagons? ☐

2 Show on the pictograph.

	1	2	3	4	5	6	7	8	9	10
△										
▮										
●										
⬡										

OXFORD UNIVERSITY PRESS

3 Use the data to finish the pictograph.

Favourite fruits in 1M

Banana	Apple	Cherry	Orange
✓ ✓ ✓ ✓ ✓ ✓	✓ ✓ ✓ ✓ ✓ ✓ ✓ ✓ ✓	✓	✓ ✓ ✓ ✓

10				
9				
8				
7				
6	🍌			
5	🍌			
4	🍌			
3	🍌			
2	🍌			
1	🍌			
	Banana	Apple	Cherry	Orange

Which fruit was the favourite?

Extended practice

a   Ask 10 people their favourite crisps flavour. Record with ticks. ✓

Plain	Salt and vinegar	Chicken	Other

b   Use the data to make a pictograph.

	1	2	3	4	5	6	7	8	9	10
Plain										
Salt and vinegar										
Chicken										
Other										

OXFORD UNIVERSITY PRESS

Favourite colours in 1T

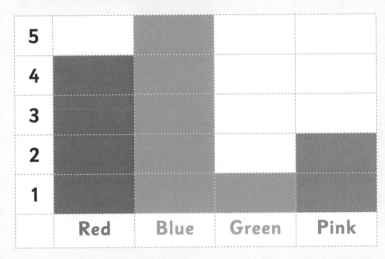

Blue is the most popular colour.

Green is the least popular colour.

Two people like **pink** best.

Four people like **red** best.

How many people are represented on the graph? How do you know?

Guided practice

1 Answer the questions.

Eye colour in 1T

a Which colour has the most?

blue	green	brown	grey

b Which has the least?

blue	green	brown	grey

c How many people have green eyes? ☐

d How many people have brown eyes? ☐

1

a Use the data in the table to make the graph.

Favourite fruit

Apple	Joe, Beth, Silo, Simon, Dom
Banana	Lee, Henry
Orange	Raj, Mason, Angela
Strawberry	Justin, Tran

Favourite fruit graph

5				
4				
3				
2				
1				
	Apple	**Banana**	**Orange**	**Strawberry**

b Which fruit is most popular?

c Which is least popular?

d How many people like strawberries best?

e How many people like bananas best?

f Which fruit does Layton like?

g How many more people like oranges than bananas?

h Who likes strawberries best?

OXFORD UNIVERSITY PRESS

Favourite subjects in Year 1

	1	2	3	4	5	6	7	8	9	10	11	12
Reading	✓	✓	✓	✓	✓	✓						
Sport	✓	✓	✓	✓	✓	✓	✓	✓	✓	✓	✓	✓
Art	✓	✓	✓									
Maths	✓	✓	✓	✓	✓	✓	✓	✓	✓			
Other	✓	✓	✓	✓	✓	✓						

2 Answer the questions.

a Which subject is most popular?

b Which is least popular?

c Which subject is the favourite of nine people?

d Which two subjects do the same number of people like?

e How many people like sport best?

f Do more people like reading or art?

What do you think "other" means?

1

a Ask 10 people what kind of pet they have and record their answers.

Cat	Dog	Fish	Other	No pet

b Make a pictograph showing the data.

Pets in our class

	1	2	3	4	5	6	7	8	9	10	11	12
Cat												
Dog												
Fish												
Other												
No pet												

c Which has the most?

cat	dog	fish	other	no pet

d Which has the least?

cat	dog	fish	other	no pet

e How many dogs?

OXFORD UNIVERSITY PRESS

Certain	Impossible	Maybe

I will go to school today.	I will dance with an alien today.	I will go to the supermarket today.

How likely is it that you will do any of these things today?

Guided practice

1 Colour the best answer.

a

certain
maybe
impossible

I will wear runners today.

b

certain
maybe
impossible

I will do mathematics today.

c

certain
maybe
impossible

I will ride a mammoth today.

1 Circle the best match.

a   This will be impossible today.

b   I will maybe go here today.

c   I will maybe eat this today.

d   This will be certain today.

OXFORD UNIVERSITY PRESS

2 Match the events with the chance of them happening today.

A cow jumps over the moon.

It starts snowing.

You see a cat on the way home.

You will travel in a car.

certain
maybe
impossible

You will leave the classroom.

You will receive a school award.

You will write a story.

Dinosaurs take over the Earth.

3 What is the chance you will pick out:

a   a red chocolate?

certain	maybe	impossible

b   a yellow chocolate?

certain	maybe	impossible

How likely is it that you will pick out a green chocolate?

Extended practice

1 Draw something:

a you will do tomorrow.	**b** you might do tomorrow.	**c** you won't do tomorrow.

- -

2 What is the chance that:

a tomorrow is a weekday?

certain	maybe	impossible

b tomorrow is the weekend?

certain	maybe	impossible

c it will rain tomorrow?

certain	maybe	impossible

d you will have pasta for dinner tonight?

certain	maybe	impossible

e you will fly to Jupiter one day?

certain	maybe	impossible

f the sun will go down later today?

certain	maybe	impossible

OXFORD UNIVERSITY PRESS

GLOSSARY

addition   The joining or adding of two numbers together to find the total. Also known as *adding*, *plus* and *sum*.

Example:

★★★ + ★★ = ★★★★★

   3 and 2 is   5

anticlockwise   Moving in the opposite direction to the hands on a clock.

area   The size of an object's surface.

Example:
It takes 12 tiles to cover this placemat.

array   An arrangement of items into even columns and rows that make them easier to count.

balance scale   Equipment that balances items of equal mass – used to compare the mass of different items. Also called pan balance or equal arm balance.

base   The bottom edge of a 2D shape or the bottom face of a 3D shape.

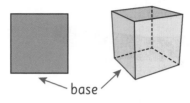

    base

calendar   A chart or table showing the days, dates, weeks and months in a year.

Month ——→ **January 2017** ←—— Year

Sun	Mon	Tues	Wed	Thur	Fri	Sat
1	2	3	4	5	6	7
8	9	10	11	12	13	14
15	16	17	18	19	20	21
22	23	24	25	26	27	28
29	30	31				

Day — (pointing to Sun/1)
Date ——→ (pointing to 15)

capacity   The amount that a container can hold.

Example:
The jug has a capacity of 4 cups.

4 cups
3 cups
2 cups
1 cup

cardinal numbers   Numbers that tell you how many things there are.

1 2 3 4 5 6

category   A group of people or things sharing the same characteristics.

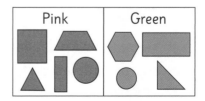

Pink	Green

centimetre A unit for measuring the length of smaller items.

Example: Length is 15 cm.

80 cm

circle A 2D shape with a continuous curved line that is always the same distance from the centre point.

clockwise Moving in the same direction as the hands on a clock.

cone A 3D shape with a circular base that tapers to a point.

corner The point where two edges of a shape or object meet.

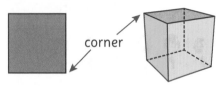

corner

cube A rectangular prism where all 8 faces are squares of equal size.

cylinder A 3D shape with 2 parallel circular bases and one curved surface.

data Information gathered through methods such as questioning, surveys or observation.

day A period of time that lasts 24 hours.

difference (between) A form of subtraction or take away.

Example: The difference between 11 and 8 is 3.

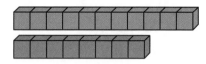

digit The single numerals from 0 to 9. They can be combined to make larger numbers.

Example: 24 is a 2-digit number.

378 is a 3-digit number.

division/dividing Sharing into equal groups.

Example: 9 divided by 3 is 3

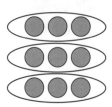

double/doubles Adding two identical numbers or multiplying a number by 2.

Example: 4 + 4 = 8   2 x 4 = 8

OXFORD UNIVERSITY PRESS

duration   How long something lasts.

Example: The school week lasts for 5 days.

edge   The side of a shape or the line where two faces of an object meet.

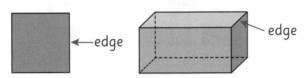

eighth   One part of a whole or group divided into eight equal parts.

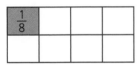

Eighth of a whole

Eighth of a group

equal   Having the same number or value.

Example:

Equal size   Equal numbers

equation   A written mathematical problem where both sides are equal.

Example: 4 + 5   =   6 + 3

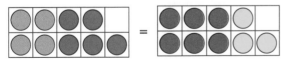

estimate   A thinking guess.

face   The flat surface of a 3D shape.

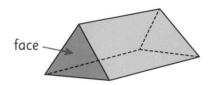

face

flip   To turn a shape over horizontally or vertically. Also known as reflection.

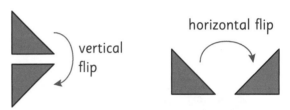

vertical flip

horizontal flip

fraction   An equal part of a whole or group.

Example: One out of two parts or $\frac{1}{2}$ is shaded.

friendly numbers   Numbers that are easier to add to or subtract from.

Example: 10, 20 or 100

half   One part of a whole or group divided into two equal parts. Also used in time for 30 minutes.

Example:

   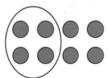

Half of a whole   Half of a group   Half past 4

hexagon   A 2D shape with 6 sides.

 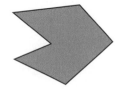

horizontal   Parallel with the horizon or going straight across.

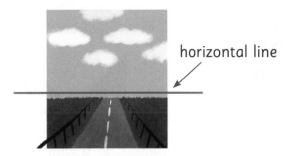

horizontal line

jump strategy   A way to solve number problems that uses place value to "jump" along a number line by hundreds, tens and ones.

Example: 16 + 22 = 38

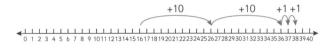

length   How long an object is from end to end.

Example: This poster is 3 pens long.

mass   How heavy an object is.

light

heavy

metre   A unit for measuring the length of larger objects.

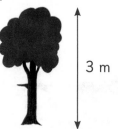

3 m

month   The time it takes the moon to orbit the Earth. There are 12 months in a year.

January   February   March
April   May   June
July   August   September
October   November   December

near doubles   A way to add two nearly identical numbers by using known doubles facts.

Example: 4 + 5 = 4 + 4 + 1 = 9

number line   A line on which numbers can be placed to show their order in our number system or to help with calculations.

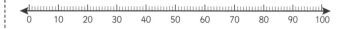

number sentence   A way to record calculations using numbers and mathematical symbols.

Example: 23 + 7 = 30

numeral   A figure or symbol used to represent a number.

Example:

1 – one   2 – two   3 – three

OXFORD UNIVERSITY PRESS

octagon   A 2D shape with 8 sides.

ordinal numbers   Numbers that show the order or position of something in relation to others.

1 2 3 4 5 6

pair   Two items that go together.

Example: Pairs that make 4

2 and 2   3 and 1

Pair of socks

parallel lines   Straight lines that are the same distance apart and so will never cross.

parallel   parallel   not parallel

partitioning   Dividing or separating an amount into parts.

Example: Some of the ways 10 can be partitioned are:

5 and 5   4 and 6   9 and 1

pattern   A repeating design or sequence of numbers.

Example: Shape pattern

Number pattern

2, 4, 6, 8, 10, 12

pentagon   A 2D shape with 5 sides.

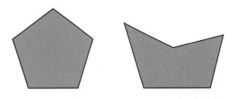

pictograph   A way of representing data using pictures to make it easy to understand.

Example: Favourite juices in our class

place value   The value of a digit depending on its place in a number.

Hundreds	Tens	Ones
		8
	8	6
8	6	3

position   Where something is in relation to other items.

Example: The boy is under the tree that is next to the house.

prism A 3D shape with parallel bases of the same shape and rectangular side faces.

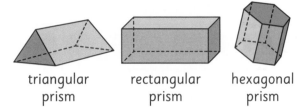

triangular prism   rectangular prism   hexagonal prism

pyramid A 3D shape with a 2D shape as a base and triangular faces meeting at a point.

square pyramid   hexagonal pyramid

quadrilateral Any 2D shape with four sides.

quarter One part of a whole or group divided into four equal parts. Also used in time for 15 minutes.

Example:

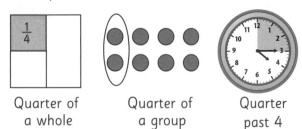

Quarter of a whole   Quarter of a group   Quarter past 4

rectangle A 2D shape with four sides and four right angles. The opposite sides are parallel and equal in length.

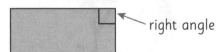

right angle

rhombus A 2D shape with four sides, all of the same length and opposite sides parallel.

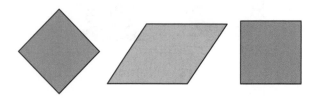

skip counting Counting forwards or backwards by the same number each time.

Example: Skip counting by 5s: 5, 10, 15, 20, 25, 30

Skip counting by 2s: 1, 3, 5, 7, 9, 11, 13

slide To move a shape to a new position without flipping or turning it. Also known as *translate*.

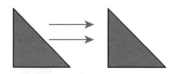

sphere A 3D shape that is perfectly round.

split strategy A way to solve number problems that involves splitting numbers up using place value to make them easier to work with.

Example: 21 + 14 = 35

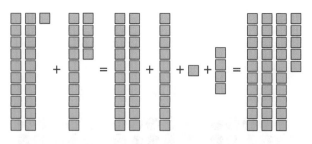

square   A 2D shape with four sides of equal length and four right angles. A square is a type of rectangle.

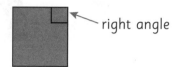
right angle

strategy   A way to solve a problem. In mathematics, you can often use more than one strategy to get the right answer.

Example: 32 + 27 = 59

Jump strategy

Split strategy

30 + 2 + 20 + 7 = 30 + 20 + 2 + 7 = 59

subtraction   The taking away of one number from another number. Also known as *subtracting*, *take away*, *difference between* and *minus*.

Example: 5 take away 2 is 3

survey   A way of collecting data or information by asking questions.

Strongly agree	☐
Agree	☑
Disagree	☐
Strongly disagree	☐

table   A way to organise information that uses columns and rows.

Flavour	Number of people
Chocolate	12
Vanilla	7
Strawberry	8

tally marks   A way of keeping count that uses single lines with every fifth line crossed to make a group.

three-dimensional or 3D   A shape that has three dimensions – length, width and depth. 3D shapes are not flat.

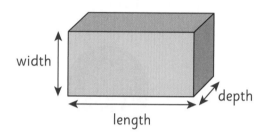
width   depth   length

trapezium   A 2D shape with four sides and only one set of parallel lines.

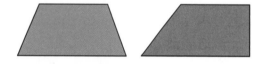

triangle   A 2D shape with three sides.

turn   Rotate around a point.

two-dimensional or 2D   A flat shape that has two dimensions – length and width.

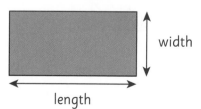
width   length

unequal   Not having the same size or value.

Example:

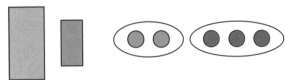

Unequal size   Unequal numbers

value   How much something is worth.

Example:

This coin is   This coin is
worth 5c.   worth $1.

vertical   At a right angle to the horizon or straight up and down.

vertical line

horizon

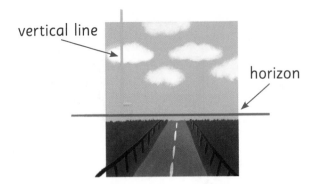

volume   How much space an object takes up.

Example: This object has a volume of 4 cubes.

week   A period of time that lasts 7 days.

Monday   Tuesday   Wednesday

Thursday   Saturday   Sunday
              Friday

whole   All of an item or group.

Example:

A whole shape   A whole group

width   How wide an object is from one side to the other.

Example: This poster is 2 pens wide.

year   The time it takes the Earth to orbit the Sun, which is approximately 365 days.

OXFORD UNIVERSITY PRESS

ANSWERS

UNIT 1: Topic 1

Guided practice

1 a **22** 23 **24**   b **36** 37 **38**
  c **54** 55 **56**   d **67** 68 **69**
  e **71** 72 **73**   f **29** 30 **31**

Independent practice

1 a 26, 27, 28, **29**, **30**, 31, **32**, **33**, **34**
  b 43, **44**, 45, **46**, **47**, **48**, 49, **50**, **51**
  c 66, 67, **68**, **69**, **70**, **71**, **72**, **73**, **74**

2 a
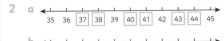

  b

  c

  d
55 54 **53** **52** 51 **50** **49** **48** **47** 46 45

Guided practice

1 a tens? 2; ones? 1; altogether? 21
  b tens? 5; ones? 3; altogether? 53
  c tens? 3; ones? 8; altogether? 38
  d tens? 6; ones? 2; altogether? 62

Independent practice

1 a

  20

  b

  50

  c

  30

2
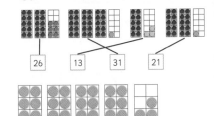

3

Extended practice

1 a 59   b 16   c 20   d 89
2 a 50   b 59   c 41   d 29

UNIT 1: Topic 2

Guided practice

1 a 12   b 28   c 15
  d 53   e 14   f 45
2 a eighteen   b forty-six

Independent practice

1

95   94   19   90

2 a seventy-one   b sixty-two
  c thirty-eight d one hundred

3
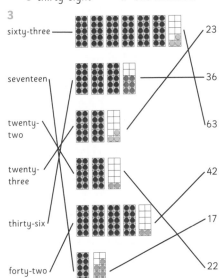

sixty-three — 23
seventeen — 36
twenty-two — 63
twenty-three — 42
thirty-six — 17
forty-two — 22

Extended practice

1 a Words: forty-five; Numeral: 45
  b Words: thirty-one; Numeral: 31
  c Words: thirteen; Numeral: 13
  d Words: seventy-seven; Numeral: 77
  e Words: one hundred and two;
    Numeral: 102

UNIT 1: Topic 3

Guided practice

1 a bigger   b bigger
2 a smaller   b smaller
  c bigger   d bigger

Independent practice

1 a 25   b 81   c 50   d 78

2
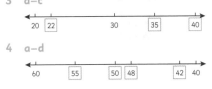

73 — More than 70
36 — 1 more than 62
63 — 1 less than 37
16 — Less than 20

3 a–c
20 22   30   35   40

4 a–d
60   55   50 48   42 40

5 13   37   48   52   67   84   113
6 105   86   74   51   39   21   15

Extended practice

1 a 725   b 18   c 143, 47   d 18, 47
2 Teacher to check. Look for answers
  that show ability to make reasonable
  estimations about where the numbers
  should go, and that space the numbers
  accurately and order the numbers
  correctly.
3 346   364   406   436   634   643

UNIT 1: Topic 4

Guided practice

1 a 12   b 17   c 17

Independent practice

1 a 19
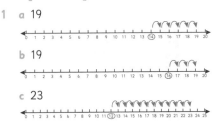

  b 19

  c 23

2 a 14. Teacher to check number line.
    Look for answers that accurately show
    the equation on the number line, using
    steps of 1, 2 or 4 to reach the total.
  b 17. Teacher to check number line.
    Look for answers that start at the
    bigger number (13) to find the answer,
    and use steps of 1, 2 or 4 to accurately
    show the solution.
  c 17. Teacher to check number line.

3 **a** 17　**b** 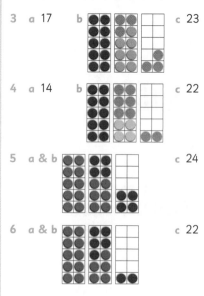　**c** 23

4 **a** 14　**b**　**c** 22

5 **a & b**　**c** 24

6 **a & b**　**c** 22

Extended practice

1 **a** 32　　**b** 31　　**c** 35
　d 44　　**e** 37　　**f** 38
2 **a** 47　**b** 60　**c** 77　**d** 94

UNIT 1: Topic 5

Guided practice

1 **a** 4 and 3　**b** 10 and 9
　c 26 is the same as 20 and 6

Independent practice

1 **a**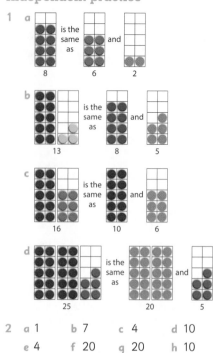
　8　　6　　2

　b
　13　　8　　5

　c
　16　　10　　6

　d
　25　　20　　5

2 **a** 1　**b** 7　**c** 4　**d** 10
　e 4　**f** 20　**g** 20　**h** 10

Extended practice

1 **a & b** Teacher to check. Look for answers that successfully identify combinations that add to the required total and that use both drawings and numbers.

---

2 **a & b** Teacher to check. Look for answers that successfully identify combinations that add to the required total and that demonstrate an understanding of place value as a basis for partitioning.

UNIT 1: Topic 6

Guided practice

1 **a** 9　　**b** 3　　**c** 12　　**d** 13

Independent practice

1 **a** 12

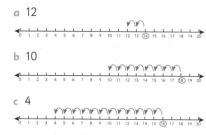

　b 10

　c 4

2 **a** 14. Teacher to check the number line. Look for answers that accurately show the equation on the number line, using steps of 1 or 2 to reach the correct answer.

　b 11. Teacher to check. Look for answers that start at the bigger number (20) to find the answer and show steps of an appropriate size (e.g. 1 or 3) to accurately reach the solution.

　c 10. Teacher to check. Look for students who start at the bigger number and count back by 1s, 2s, 5s or 10s to find the correct answer.

3 **a** 15　**b** 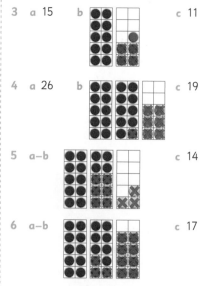　**c** 11

4 **a** 26　**b**　**c** 19

5 **a–b**　**c** 14

6 **a–b**　**c** 17

NOTE: for questions 3–6 the specific counters crossed out are not important, as long as the correct number has been crossed out.

Extended practice

1 **a** 7　　**b** 15　　**c** 22　　**d** 22
2 **a** 32　**b** 38　**c** 57　**d** 75
　e 27　**f** 86

UNIT 1: Topic 7

Guided practice

1 **a** 4　　　　**b** 7　　　　**c** 4
　d 2　　　　**e** 5　　　　**f** 3

Independent practice

1 **a** 4 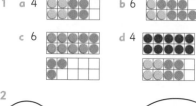　**b** 6

　c 6　　　　　　**d** 4

2
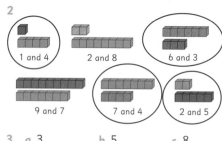
1 and 4　　2 and 8　　6 and 3
9 and 7　　7 and 4　　2 and 5

3 **a** 3　　**b** 5　　**c** 8
4 **a** 8　　**b** 6　　**c** 8

Extended practice

1 Note: pairs can be in any order
　7 and 11　　25 and 21　　18 and 14
　19 and 23　　16 and 20

2 **a** 4　　　**b** 6　　　**c** 9
Teacher to check number lines. Look for answers that show ability to use efficient strategies such as jumping by 2 and that accurately show working using the number line.

UNIT 1: Topic 8

Guided practice

1 **a** 2, 4, 6, 8, **10, 12, 14, 16, 18, 20, 22**

　b 5, 10, 15, 20, **25, 30, 35, 40, 45, 50, 55**

　c 10, 20, 30, **40, 50**

Independent practice

1 **a**

　b

　c

2 a 38, 40, 42, **44**, 46, **48**, 50, **52**, **54**, 56

b 35, 40, **45**, 50, **55**, **60**, 65, **70**, 75, **80**

c 10, **20**, 30, **40**, **50**, **60**, 70, **80**, **90**, 100

3 a 5, 10, 15, 20, 25

b 10, 20, 30, 40, 50

c 2, 4, 6, 8, 10, 12, 14

d 5, 10, 15, 20, 25, 30, 35, 40, 45, 50, 55, 60

Extended practice

1

73	88	66	98	65	56	100	98
68	87	86	28	72	70	88	96
76	78	80	82	84	48	60	94
74	72	48	90	86	88	90	92
71	70	63	78	68	46	64	72

2

26	14	64	46	49	52	33	78	84	3
41	5	80	65	44	30	94	22	17	63
53	37	28	10	12	15	16	75	39	81
92	56	70	35	86	60	95	50	20	47
93	87	32	55	94	91	6	25	87	59
39	45	40	85	27	21	73	90	99	77
32	24	63	72	58	68	66	43	51	31

Secret number: 34

UNIT 1: Topic 9

Guided practice

1 a 4   **b** 3

Independent practice

1 a

b 5

2 a

b 2

---

3 a

b 2

4 a 12 shared between 3 is 4.

b 8 shared between 4 is 2.

c 12 shared between 6 is 2.

d 15 shared between 3 is 5.

Extended practice

1 a Teacher to check. Look for answers that show ability to successfully represent 12 items and that demonstrate an understanding of equality by dividing the total into three equal groups.

b 4

2 a Teacher to check. Look for answers that show ability to successfully represent 15 items and that demonstrate an understanding of equality by dividing the total into five equal groups.

b 15 shared between 5 is 3.

UNIT 1: Topic 10

Guided practice

1 Teacher to check.

2 a red   **b** grey   **c** green

Independent practice

1 Teacher to check.

2 Teacher to check.

3

4 first, second, third, fourth

5 a cat   **b** cow   **c** dog   **d** frog

6 a

b

c

Extended practice

1 Teacher to check.

2

□	☺	○	🌳		△	☺
1	2	3	4	5	6	7

UNIT 2: Topic 1

Guided practice

1 a 2   **b** 4

2 quarters

3 half

Independent practice

1

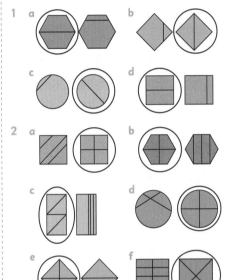

2

3 a–c Teacher to check. Look for answers where the shapes have been divided into two pieces and where the pieces are of approximately the same size.

4 a–c Teacher to check. Look for answers where the shapes have been divided into four pieces and the fractions look to be of approximately equal size.

5

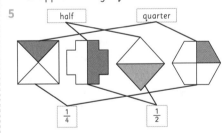

Extended practice

1 Teacher to check. Look for answers that show more than one solution and whose four parts are of approximately equal size.

2 NOTE: the particular segments coloured is unimportant.

a   b

c   d

UNIT 2: Topic 2

Guided practice

1 **a** halves   **b** quarters
   c quarters   **d** halves

Independent practice

1 **a** Two groups of 7 frogs should be circled.
   b 2   **c** 7

2 **a** Four groups of 4 apples should be circled.
   b 4   **c** 4

3

4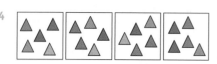

5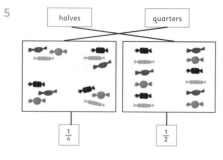

Extended practice

1 **a** 24
   b 12 circles should be coloured red.
   c 6 circles should be coloured blue.
   d 12   **e** 6   **f** half
   g quarters   **h** quarter

UNIT 3: Topic 1

Guided practice

1

2 **a** $2, $1, 50c, 20c, 10c, 5c
   b $1 and $2   **c** 5c

Independent practice

1 **a**
   b 20c, 50c   **c** $2

2 **a**

   b

3 **a**

   b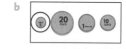

4 **a**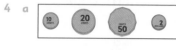

   b

5

Extended practice

1 **a** 6   **b** 30c
2 **a** 4   **b** $2
3 **a** 5   **b** $10
4 **a** 30c   **b** $3

UNIT 4: Topic 1

Guided practice

1 **a**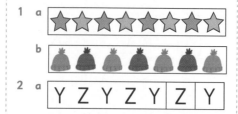

   b

2 **a** Y Z Y Z Y Z Y

   b A B C A B C A

Independent practice

1 **a** 3   **b** 2   **c** 2   **d** 3

2 **a**

   b Teacher to check.

3 **a**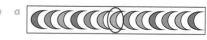

   b Teacher to check.

4 **a**

   b pink

5 **a** B B A B B A A B A

   b B

Extended practice

1 **a**

   b

2 **a & b** Teacher to check. Look for answers that demonstrate an understanding of repeating colour patterns and that accurately describe the pattern created.

3 **a & b** Teacher to check. Look for answers that demonstrate an understanding of either a repeating or a growing pattern using shapes and that accurately describe the pattern created.

UNIT 4: Topic 2

Guided practice

1 **a**

1	②	3	④	5	⑥	7	⑧	9	⑩
11	⑫	13	⑭	15	⑯	17	⑱	19	⑳
21	㉒	23	㉔	25	㉖	27	㉘	29	㉚

   b 2, 4, 6, 8, 0 (sequence can start at any point, e.g. 0, 2, 4, 6, 8)

   c 32, **34**, **36**, **38**, **40**

Independent practice

1 **a**

1	2	3	4	⑤	6	7	8	9	⑩
11	12	13	14	⑮	16	17	18	19	⑳
21	22	23	24	㉕	26	27	28	29	㉚
31	32	33	34	㉟	36	37	38	39	㊵
41	42	43	44	㊺	46	47	48	49	㊿

   b 5, 0 (in any order)

OXFORD UNIVERSITY PRESS

c 55, **60**, **65**, **70**

2 a

1	2	3	4	5	6	7	8	9	10
11	12	13	14	15	16	17	18	19	20
21	22	23	24	25	26	27	28	29	30
31	32	33	34	35	36	37	38	39	40
41	42	43	44	45	46	47	48	49	50
51	52	53	54	55	56	57	58	59	60
61	62	63	64	65	66	67	68	69	70
71	72	73	74	75	76	77	78	79	80
81	82	83	84	85	86	87	88	89	90
91	92	93	94	95	96	97	98	99	100

b 0   c even

3 a & c

1	2	3	4	5	6	7	8	9	10
11	12	13	14	15	16	17	18	19	20
21	22	23	24	25	26	27	28	29	30
31	32	33	34	35	36	37	38	39	40
41	42	43	44	45	46	47	48	49	50
51	52	53	54	55	56	57	58	59	60
61	62	63	64	65	66	67	68	69	70
71	72	73	74	75	76	77	78	79	80
81	82	83	84	85	86	87	88	89	90
91	92	93	94	95	96	97	98	99	100

b 19   d 19

4 a 35, 40, 45, **50**, **55**, **60**, 65, **70**, **75**, 80. Counting by? 5

b 40, 50, **60**, 70, **80**, **90**, 100. Counting by? 10

c 20, 22, 24, **26**, 28, **30**, **32**, **34**, 36, **38**. Counting by? 2

Extended practice

1 a–c &   2 a–c

1	2	3	4	5	6	7	8	9	10
11	12	13	14	15	16	17	18	19	20
21	22	23	24	25	26	27	28	29	30
31	32	33	34	35	36	37	38	39	40
41	42	43	44	45	46	47	48	49	50
51	52	53	54	55	56	57	58	59	60
61	62	63	64	65	66	67	68	69	70
71	72	73	74	75	76	77	78	79	80
81	82	83	84	85	86	87	88	89	90
91	92	93	94	95	96	97	98	99	100

1 d 8

2 d 4, 5, 9

3 a 100, 102, **104**, **106**, 108, 110, **112**, **114**, **116**, **118**

b 105, 110, 115, **120**, 125, **130**, **135**, 140, **145**, **150**

c 100, **110**, 120, 130, **140**, **150**, 160, **170**, **180**, **190**

UNIT 5: Topic 1

Guided practice

1 a 6 paperclips long
  b 9 paperclips long
  c 5 paperclips long

2 zucchini

Independent practice

1 a–c Teacher to check. Look for reasonable estimates of the length of the items in paperclips and answers that have been measured accurately by placing the paperclips end to end with no gaps.

  d Approximately 5 small or 4 large paperclips long.

2 The pencil is likely to be the shortest item. Look for answers that include reasoning, using language of measurement such as shorter and longer.

Guided practice

1 a 8 tiles   b 24 tiles   c 4 tiles

2 calendar

Independent practice

1 a–d Teacher to check. Look for reasonable estimates of the area of the items, taking into account the size of the block or tile being used, and for answers that demonstrate an ability to accurately measure by placing the tiles or blocks with no gaps.

2 Teacher to check; most likely to be the book or the lunch box lid. Look for answers that include reasoning and that demonstrate an understanding of the concept of area.

Extended practice

1 a–b Teacher to check. Look for answers that demonstrate accurate measurement techniques, placing the items end-to-end with no gaps or overlaps.

2 pencils

3 a–b Teacher to check. Look for answers that demonstrate accurate measurement techniques, placing units in rows with no gaps or overlaps.

4 Teacher to check – answers will vary depending on the size of the blocks and sticky notes used. Look for answers that include reasoning using the language of measurement.

5 a–b Teacher to check. Look for answers that demonstrate that students can competently compare the area of two different objects and can accurately measure using informal units.

6 Teacher to check. Answers will vary depending on the size of the blocks and the sticky notes used. Look for answers that include reasoning using the language of measurement.

UNIT 5: Topic 2

Guided practice

1 a 3 cubes   b 6 cubes
  c 9 cubes   d 7 cubes

Independent practice

1 a 6 cubes   b 4 cubes
  c 12 cubes   d 9 cubes

  Teacher to check students' models. Look for responses that accurately make the model using cubes and that can use the physical model to identify the volume.

2 a Model C should be circled in blue.
  b Model B should be circled in red.

3 a B   b D

Guided practice

1 a 4 cups   b 6 cups
  c 10 cups   d 8 cups

Independent practice

1 a spoon   b mug
  c mug   d bucket

  There may be an opportunity to discuss the concept of the most appropriate units to use as students respond to this question. For example, it is possible to measure the capacity of the fish tank using the coffee mug but it is not the quickest or most efficient way of doing it.

2 a–b Teacher to check. Look for reasonable estimates of items that have a greater and smaller capacity than the saucepan and justification of answers using the language of capacity.

  c Answers will vary depending on the items drawn in a & b. Most likely the mug or bucket will be appropriate for the first item and the spoon or mug for the second. Look for answers that provide justification and that demonstrate an understanding of how to choose the most appropriate unit.

Extended practice

1 **a–b** Teacher to check. Look for students who are able to construct two different models with a volume of 8 cubes, and who can describe their models using the language of volume.

2 **a–b** Teacher to check. Look for students who are able to make reasonable estimates of the capacity of their chosen containers in cups, and who are then able to accurately measure and record the results.

UNIT 5: Topic 3

Guided practice

1 **a–b** Teacher to check. Look for answers that show an understanding of the concepts of lighter and heavier and that demonstrate reasonable choices in comparison with the items shown – e.g. a glue stick would be lighter than the paint can and a pencil would be lighter than the calculator.

2 **a–b** Teacher to check. Look for answers that show an understanding of the concepts of lighter and heavier and that demonstrate reasonable choices in comparison with the items shown – e.g. a bottle of water would be heavier than the cupcake and a car would be heavier than the pumpkin.

Independent practice

1 **a–d** Teacher to check. Look for answers that show ability to choose pairs with an obvious difference in mass, and to put the heavier and lighter item in each pair on the correct side of the pan balance.

2 Answers will vary depending on the mass of each student's pencil case and the versions of the items chosen. Look for answers that show ability to use strategies such as hefting to accurately predict the results and ability to correctly use a pan balance to check.

Likely results are:

a lighter   b heavier   c heavier
d lighter   e heavier   f lighter

Extended practice

1 **a–d** Answers will vary depending on the size of the cubes and counters used. Look for answers that demonstrate ability to achieve a reasonable balance between the given number of cubes and the required number of counters and that demonstrate an understanding of equality of mass.

2 teabag, teaspoon, coffee mug, milk, kettle

Accept slight variances if students can justify their responses – e.g. the kettle may be lighter than the milk container if it is empty.

UNIT 5: Topic 4

Guided practice

1 a o'clock   b o'clock   c half past
  d half past   e o'clock   f half past

2 a 5 o'clock   b half past 8   c half past 3

Independent practice

1
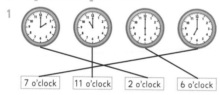

| 7 o'clock | 11 o'clock | 2 o'clock | 6 o'clock |

2 a   b

c   d

e   f

3
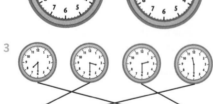

| half past 3 | half past 11 | half past 7 | half past 2 |

4 a   b

c   d

e   f

Extended practice

1
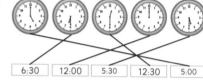

| 6:30 | 12:00 | 5:30 | 12:30 | 5:00 |

2 a 8:00   b 7:30   c 6:00
  d 3:30   e 11:30   f 11:00

3 half past four, 4:30

UNIT 5: Topic 5

Guided practice

1 a months   b hours   c days
  d weeks   e hours   f months

Independent practice

1 **a–b** Teacher to check. Look for answers that demonstrate an understanding of duration by drawing from familiar events to choose options that take longer than the given times, and that use the language of time to justify responses.

2 a

| 5 days | 4 weeks | 2 hours | 4 months |

  b Watching a movie.

3 a 2   3   1
  b 1   2   3
  c 3   2   1

Answers may vary depending on when the student's birthday is.

4 Answers will vary depending on when the student's birthday is. Look for answers that justify the response using the language of duration.

Extended practice

1 a 24   b 7   c 4   d 12

2 **a–f** Answers will vary. Look for answers that identify appropriate units to measure time, for example, hours for shorter time periods such as the time until dinner, and weeks for longer periods such as the time until the end of term.

OXFORD UNIVERSITY PRESS

3 **a–b** Teacher to check. Look for answers that demonstrate an understanding of the relative duration of events, and for plausible estimates of the duration of activities chosen by the students.

UNIT 6: Topic 1

Guided practice

1 **a** 2 horizontal lines, 2 vertical lines, 4 corners, 4 sides

b 2 horizontal lines, 0 vertical lines, 6 corners, 6 sides

c 1 horizontal line, 1 vertical line, 3 corners, 3 sides

d 1 horizontal line, 2 vertical lines, 5 corners, 5 sides

Independent practice

1 **a & b**

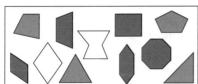

2

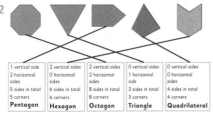

1 vertical side 2 horizontal sides 5 sides in total 5 corners **Pentagon**	2 vertical sides 0 horizontal sides 6 sides in total 6 corners **Hexagon**	2 vertical sides 2 horizontal sides 8 sides in total 8 corners **Octagon**	0 vertical sides 1 horizontal side 3 sides in total 3 corners **Triangle**	0 vertical sides 0 horizontal sides 4 sides in total 4 corners **Quadrilateral**

3 **a** 4 **b** 3 **c** 4
  d 5 **e** 1

4 **a** parallel **b** not parallel
  c parallel **d** not parallel
  e parallel **f** not parallel

Extended practice

1 **a–b** Teacher to check. Look for answers that show ability to draw a shape that meets the criteria, and that demonstrate an understanding of the key language.

2 **a** hexagon **b** octagon

Teacher to check the descriptions. Look for answers that show ability to use the language of shape, including sides, corners and line types, to accurately describe the shapes.

UNIT 6: Topic 2

Guided practice

1

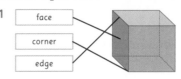

2 cylinder, cone

Independent practice

1 The following objects should be circled:
  a cube **b** triangular prism
  c sphere **d** triangular prism
  e sphere **f** triangular prism

2 The cone, sphere and cylinder should be circled.

3 **a** 3 **b** 2 **c** 4
  d 2 **e** 3

4

Extended practice

1 **a** cube **b** cylinder

2 **a** drawing of a rectangle
  b drawing of a rectangle
  c drawing of a square or a smaller rectangle that shows the proportion of the side view

UNIT 7: Topic 1

Guided practice

1 **a** in the tree **b** in the shed
  c next to the tree **d** under the car

Independent practice

1 **a–e** Teacher to check. Look for answers that show ability to accurately interpret positional language to correctly place the items.

2 NOTE: accept either written or drawn answers from students.
  a the train **b** the boat
  c the blocks **d** the duck

3 **a & b** Answers will vary. Look for answers that show ability to accurately use positional language such as above, next to, left of, etc. to describe the position of each item.

Extended practice

1 **a–b** Answers will vary. Look for answers that show ability to accurately use positional language such as above, below, near, between etc. to describe the position of each item.

2 **a–b** Teacher to check. Look for answers that show an understanding of positional language in describing where the dog is.

UNIT 7: Topic 2

Guided practice

1 **a** clockwise **b** anticlockwise
  c anticlockwise **d** clockwise

2 **a** backwards **b** forwards

Independent practice

1 **a** clockwise **b** anticlockwise

2 **a** anticlockwise **b** clockwise

3 **a** clockwise **b** anticlockwise

4 **a–g**

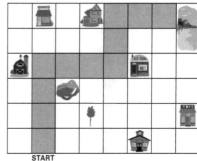

START

h the beach

Extended practice

1 **a & b** Teacher to check. Look for answers that show ability to accurately use language such as left, right, forwards, backwards, clockwise and anticlockwise to accurately describe the paths. Likely responses:

a Move forward 2 spaces. Turn right. Move forward 1 space.

b Move forward 3 spaces. Turn right. Move forward 3 spaces. Turn left. Move forward 1 space.

UNIT 8: Topic 1

Guided practice

1

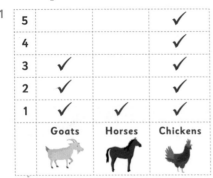

5			✓
4			✓
3	✓		✓
2	✓		✓
1	✓	✓	✓
	Goats	Horses	Chickens

Independent practice

1 a 7   b 5   c 4   d 9

2

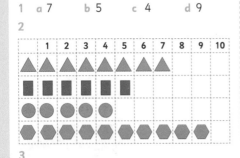

	1	2	3	4	5	6	7	8	9	10
▲	▲	▲	▲	▲	▲	▲	▲			
■	■	■	■	■	■					
●	●	●	●	●						
⬡	⬡	⬡	⬡	⬡	⬡	⬡	⬡	⬡	⬡	

3

10		🍎		
9		🍎		
8		🍎		
7		🍎		
6	🍌	🍎		
5	🍌	🍎		🍊
4	🍌	🍎		🍊
3	🍌	🍎		🍊
2	🍌	🍎		🍊
1	🍌	🍎	🍒	🍊
	Banana	Apple	Cherry	Orange

Extended practice

1 a Answers will vary. Look for answers that show ability to accurately record the responses of 10 students in the table using ticks or tally marks.

b Answers will vary. Look for answers that show ability to use the data from the previous question to make an accurate pictograph using one-to-one correspondence.

UNIT 8: Topic 2

Guided practice

1 a brown b grey   c 2   d 5

Independent practice

1 a

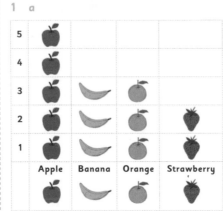

5	🍎			
4	🍎			
3	🍎	🍌	🍊	
2	🍎	🍌	🍊	🍓
1	🍎	🍌	🍊	🍓
	Apple	Banana	Orange	Strawberry
	🍎	🍌	🍊	🍓

  b apple   c strawberry
  d 2   e 3   f banana
  g 1   h Justin and Tran

2 a sport   b art   c maths
  d reading and other   e 12
  f reading

Extended practice

1 a Answers will vary. Look for answers that show ability to accurately record classmates' responses in the table. Note that in some instances the total responses might be more than 10 if some students surveyed have more than one pet.

b Responses will vary depending on data collected. Look for answers that demonstrate ability to accurately represent the data in a pictograph.

c–e Responses will vary depending on the data collected. Check that the answer accurately interprets the data.

UNIT 9: Topic 1

Guided practice

1 a Teacher to check. Look for answers that include justification using the language of chance.

  b Answers will vary depending on individual class timetable.

  c impossible

Independent practice

1 a–d Answers may vary based on students' experiences and situations. The most likely responses are below; however, any plausible response should be accepted if the student can give adequate reasoning.

  a Child flying a plane should be circled.

  b Child at the supermarket and/or cinema should be circled.

  c The sandwich or bowl of cereal should be circled.

  d Child having a drink should be circled.

2 Answers will depend on students' individual circumstances. Look for answers that show ability to correctly categorise impossible events, such as dinosaurs taking over the Earth, and that offer plausible explanations for their choices.

3 a maybe   b impossible

Extended practice

1 a Teacher to check. Look for answers that offer plausible choices for each likelihood category and that can justify reasoning using the language of chance.

2 a Certain or impossible, depending on the current day.

  b Certain or impossible, depending on the current day.

  c Maybe

  d Could be any, depending on the student's reasoning.

  e Impossible

  f Certain

OXFORD UNIVERSITY PRESS